OPEN-ENDED QUESTIONS IN ELEMENTARY MATHEMATICS: INSTRUCTION & ASSESSMENT

Mary Kay Dyer

Christine Moynihan

EYE ON EDUCATION

6 DEPOT WAY WEST, SUITE 106

LARCHMONT, NY 10538

(914) 833–0551

(914) 833–0761 fax

www.eyeoneducation.com

Library of Congress Cataloging-in-Publication Data

Dyer, Mary Kay 1946–
    Open-ended questions in elementary mathematics : instruction and assessment / by Mary Kay Dyer and Christine Moynihan.
    p. cm.
    Includes bibliographical references.
    ISBN 1-935055-60-04
    1. Mathematics—Study and teaching (Elementary) I. Moynihan, Christine, 1951– II. Title.

    QA135.5 .D98 2000
    372.7—dc21

                                                                    00--028790

10 9 8 7 6 5 4 3 2

Cover design and art by Carolyn H. Edlund
Editorial and production services provided by
Richard H. Adin Freelance Editorial Services
52 Oakwood Blvd., Poughkeepsie, NY 12603-4112
(914-471-3566)

MEET THE AUTHORS

Mary Kay Dyer has taught mathematics at most grade levels and has also been an instructor at several Boston area colleges. She has completed doctoral level courses at Boston University. After many years as the coordinator of mathematics for Newton, Massachusetts, Mary Kay has retired from public education but is consulting for several school districts and collaboratives in New England. She has been very involved in professional education organizations including speaking at meetings, chairing the 1995 annual meeting of the National Council of Teachers of Mathematics, and acting as NCTM's northeast regional representative for three years. Other books co-authored by Mary Kay are *Power Polygons* (Cuisenaire/Dale Seymour), *Geopiece Math* (Delta Education) and *Using Computers in Mathematics Education* (Addison Wesley).

Christine Moynihan has been a mathematics curriculum specialist for eight years. She is currently the mathematics and science specialist in Wayland, Massachusetts. Prior to specializing in curriculum, Chris spent seventeen years as a classroom teacher in kindergarten and grades two, three, five and six. She holds a Ph.D. in curriculum and instruction from Boston College. Chris consults with school systems, privately tutors and presents at local, regional, and national conferences. Special interests include helping teachers deepen and refine their assessment practices as well as improving the mathematics opportunities for girls which she does by running a girls-only summer mathematics camp.

TABLE OF CONTENTS

1

WHY USE OPEN-ENDED QUESTIONS?

The role of the elementary classroom teacher is a multifaceted one. What an understatement! Elementary classroom teachers are expected to be experts on the content knowledge, instructional methods, and assessment practices in reading, language arts, mathematics, science, and social studies. In addition, the elementary teacher must also be well-versed in areas such as the arts, technology, health, sex education, and cultural diversity, to name just a few. Further, the elementary teacher fills the roles of psychologist, mediator, evaluator, diagnostician, and nurturer as well as a host of others. It is no wonder elementary teachers often feel overwhelmed by the range of responsibilities that they consistently face in meeting the day-to-day challenges of life in the classroom.

Why, then, should elementary educators be interested in adding to an already full plate by using open-ended questions in all content areas and especially in mathematics?

♦ Will open-ended questions help them answer the calls for curricular reform?

♦ Will open-ended questions support their instruction in ways aligned with research-based views of learning and teaching?

♦ Will open-ended questions impact their assessment practice in ways that give them entree into each student's understanding of concepts, ability to use processes, and ability to communicate?

The answer to the preceding questions is a resounding "Yes!"

1

CALLS FOR CURRICULAR REFORM

NEEDS OF THE WORKPLACE AND SOCIETY IN THE NEW CENTURY

The decade of the 1980s was littered with reports that sought to bring into focus what is needed to help us prepare for life in the twenty-first century. *A Nation at Risk* (National Commission on Excellence in Education, 1983) and *Educating Americans for the Twenty-first Century* (National Science Board Commission, 1983) are credited with alerting the country to the fact that students were not prepared for work in the current workplace much less for the workplace of the future. Other studies and reports—*The Underachieving Curriculum* (McKnight et al., 1987), *Results from the Fourth National Assessment of Educational Progress* (Lindquist, 1988), *Everybody Counts* (National Research Council, 1989), and *Reshaping School Mathematics* (Mathematical Sciences Education Board, 1990)—also pointed out the same sources of concern: student performance, the needs of business and industry, and the mismatch between them.

In 1987, Henry Pollak, an industrial mathematician, strengthened the call for reform in mathematics education by describing what is expected of employees as including the ability to:

♦ Structure problems based on an understanding of their mathematical features

♦ Utilize multiple techniques in problem solving

♦ Transfer and apply mathematical ideas

♦ Work toward the solution of "open" problems

♦ Work with others

The U.S. Congressional Office of Technology Assessment supported that view in 1988 by stating that employees must be prepared to handle the massive amounts of information with which they will be presented within a complex, technology-laden context in a cooperative work setting. This did much to highlight the difference between what is needed as we head into the next century and the bookkeeping arithmetic still at the core of most elementary programs.

NATIONAL RESPONSE

New goals for society, the workplace, and students were identified in these reports and others as a base from which to build restructuring efforts. These goals were clarified and supported by the release of the National Council of Teachers of Mathematics' *Curriculum and Evaluation Standards for School Mathematics* (NCTM, 1989), which was offered as a framework for the curricular restructuring needed in mathematics—a guide for the work to take place in the decade of the 1990s. It is a document that took years to complete and is remarkable in that it represents a consensus about what can and should be done to improve the mathematics education for all children. The bottom-line message of the *Standards* is that all children deserve a quality mathematics education that is broader in scope in terms of content, more attentive to the needs of students as learners, more aligned with assessment practices that reflect what students know and understand.

An update of the *Standards* called *Principles and Standards for School Mathematics* (NCTM, 2000) incorporates feedback to the original document, new research, and technological advances that have taken place since 1989. While there have been changes, much of the original thinking of the *Standards* has been preserved. The new document still espouses the premise that all students are deserving of a rigorous, high-quality mathematics education. It describes standards but goes a bit further in identifying six principles for mathematics instructional programs:

- ◆ Equity: Mathematics instructional programs should promote the learning of mathematics by all students.

- ◆ Mathematics curriculum: Mathematics instructional programs should emphasize important and meaningful mathematics through curricula that are coherent and comprehensive.

- ◆ Teaching: Mathematics instructional programs depend on competent and caring teachers who teach all students to understand and use mathematics.

- ◆ Learning: Mathematics instructional programs should enable all students to understand and use mathematics.

- ◆ Assessment: Mathematics instructional programs should include assessment to monitor, enhance, and evaluate the mathematics learning of all students and to inform teaching.

- ◆ Technology: Mathematics instructional programs should use technology to help all students understand mathematics and prepare them to use mathematics in an increasingly technological world (NCTM, 1998).

STATE AND LOCAL RESPONSES

The mathematics curriculum reform movement moved ahead at the state level in the nineties as the *Standards* was used as a guide for making changes in curriculum, instruction, and assessment. Many states used the document as the basis for identifying their mathematical needs at a more local level set within the larger context of the national calls for reform. In Massachusetts, for example, an outgrowth of the Education Reform Act of 1993 was the writing of a curriculum framework in each of seven content areas. The authors of the Massachusetts Mathematics Curriculum Framework Development Committee (1997) used the *Standards* as the foundation for their work as they outlined core concepts, habits of mind, and learning standards that align with it so that meaningful and sustained curricular reform can be effected.

How can the use of open-ended questions help to answer the calls for curricular reform? Quite simply, open-ended questions enable teachers to see more clearly what all of their students know and understand. Such questions allow each child to respond at his/her level in a way that is meaningful. This can give teachers a window into students' thinking and sometimes even open a door! When a child is asked "How did you figure that out?" after solving 26 + 18, for example, the child's conceptual understanding of the operation of addition can be made apparent by his or her description of the process. With that type of information, teachers are more capable of making instructional decisions that are aligned with their students' needs. Further, when students are asked to explain their thinking as they often are in open-ended questions, they are refining and deepening their understandings. Use of open-ended questions can help teachers realize the primary thrust of mathematics curricular reform—that all students know and understand mathematics.

CHANGES IN TEACHING AND LEARNING

NEW VIEWS OF LEARNING

When recommendations are made that directly affect teaching and instruction, the impetus usually comes from new views of learning. The major question focuses on answering, "How do students learn best—not just some students, but *all* students?" Since the *Standards*, that phrase "all students" has become very significant. It implies that each and every student regardless of gender, race, ethnicity, socioeconomic class, learning style, and level of performance is deserving of equal opportunity to experience a rich, meaningful, relevant, challenging mathematics education so that she/he understands, knows, and can do mathematics. With open-ended questions, it is possible to nurture multiple approaches to problem solving rather than forcing all children to respond alike. Such divergent thinking is likely what will be needed to deal with the societal demands of the twenty-first century.

Another basic tenet that serves as a foundation for the *Standards* and the reform curricula developed during the 1990s is that to "know" mathematics means to "do" mathematics. This notion is hardly new, however, since it can be traced back to John Dewey (1916) and his then-radical, hands-on approach to learning. His belief rests on the relatively simple assumption that prior knowledge is used as a base for new learning and knowledge. This view of learning is today referred to as constructivism. Recent research confirms that children do not really "learn" by being passive receivers of information. They, instead, learn when they actively construct their own knowledge as a result of personal experiences that they assimilate in conjunction with prior learnings thereby constructing new meaning and knowledge.

Encouraging students to become lifelong independent learners of mathematics seems critical when the following is considered. Mathematics can no longer be thought of as a finite collection of rules, algorithms, and procedures. In the early 1980s, it was estimated that half of what was known in the field of mathematics had been "invented" or "discovered" since World War II (Davis & Hersh, 1981). It was predicted in the late 1980s that by the new millennium this would double (Lindquist, 1989). It will not be sufficient, therefore, to be a good memorizer of already existing rules and procedures. What is and will continue to be needed is the ability to ask and answer the whys and hows—Why did you try this? Why does this work? How does this work? How did your prior knowledge connect with what you have learned?

NEW VIEWS OF TEACHING

Open-ended questions support the needed shift from viewing mathematics as a finite collection of rules and almost force a change in instructional practice. Their use dramatically impacts what still stands today as the predominant view of how to teach mathematics. That view may sound familiar to most of today's adults as the way that mathematics instruction happened then and still happens now. The teacher begins class by asking students to trade the previous night's homework papers with one another so that the correct answers can be read. If many people get one wrong, the teacher may either ask a student to solve the problem on the board or do so her/himself. The next phase of the lesson is the teacher introducing the new material by showing a few examples on the board and leading the class through the steps. There may be some variation in what follows, for example, a page full of similar problems to be completed individually for classwork or the chance to begin that night's homework, which could be anywhere from 10 to 50 practice problems.

In contrast, when a teacher presents an open-ended question to students, she/he does so with a teaching style far different from the one just described. Students are invited to reflect more deeply about a single question or problem that is presented at the beginning of the class period. For example, students in grade four may be asked, "Which is larger: one-fourth or one-seventh, and how do you know?" They are invited to "muck around" and investigate the question in class, most often with a partner or even in a small group. After some time, students are asked to share their thinking, processes, and solutions with the class. The teacher may then summarize the class findings and assign homework that is reflective of what happened in class in both style and substance.

How, then, does the use of open-ended questions help teachers integrate new views of learning and expand their repertoires for teaching? Open-ended questions ask students to *do* mathematics—by requiring them to work out a solution path or paths, explain their thinking and reasoning, outline their strategy process, and discuss how they used what they already knew to find out what they did not know. The focus is shifted to the learner to think and do, do and think, and explain and justify. The teacher neither serves as the dispenser of knowledge nor stands as the only one who explains procedures and processes. The teacher's role is to facilitate students work with challenging questions that are interesting, mathematically significant, and allow all students entry into the task.

CHANGES IN ASSESSMENT

NEW GOALS

Traditionally, much of the planning for curriculum, instruction, and assessment occurs in a somewhat linear fashion. Most people begin with the curriculum component by examining just what is being taught primarily in terms of specific procedures and skills and, perhaps, secondarily, in terms of concepts. Once that has been identified, the next step is to think about how to go about teaching all those procedures, skills, and concepts—the instruction component. This step is followed by thinking about how to assess if the students have learned what has been taught—the last of the three components—the assessment component.

This way of thinking about assessment rests on the belief that assessment is primarily about assigning a grade of some sort to each student. A broader and newer view of assessment, however, is one that grows from the definition of assessment put forth by the NCTM in *Assessment Standards in School Mathematics* (1995) that sees assessment as "the process of gathering evidence of a student's knowledge of, ability to use, and disposition toward mathematics, and of making inferences from that evidence for a variety of purposes." This NCTM document was designed to expand upon and complement the *Standards* and was based upon extensive research, in particular, the report *Measuring What Counts* from the Mathematical Sciences Education Board (1993).

This expanded view of assessment frames four major reasons for assessment with their correlating results:

♦ *Monitoring students' progress* can promote continued growth toward both individual and collective goals.

♦ *Making instructional decisions* based upon assessment of students can help teachers modify and improve instruction.

♦ *Evaluating students' achievement* at regular intervals as a result of evidence received from multiple sources can help teachers recognize the accomplishments of students.

♦ *Evaluating instructional programs* can lead to modifications that aim to encourage all students to meet high expectations.

NEW APPROACHES

How can the use of open-ended questions address and support this wider view of assessment? Open-ended questions can help students share their knowledge of mathematics in ways that traditional multiple-choice, fill-in-the-blank, true/false, just-give-the-right-answer types of assessments cannot. Those assessments can give some information about procedural skills but yield little or no information about conceptual understanding or knowledge and the ability to apply it in new situations. When students are asked to show their work and explain their thinking as they often are in open-ended questions, they are given the opportunity to show what they know and understand. As a result, teachers have a clearer idea of what students truly understand about a mathematical concept or idea as opposed to what they may have just memorized as steps to a procedure. For example, the child who explains that 26 + 18 can be solved by adding 26 and 20 and then subtracting 2 is providing the teacher with information about what he/she knows about rounding to a familiar, friendly number and then compensating for that change. This gives the teacher a much clearer idea of what the child knows about numbers and their relationships as opposed to having the child write the traditional algorithm on paper that involves "putting down the 4 and carrying the 1" with no supporting detail of thought.

Teachers are often frustrated and perplexed when students do not do well on word problems when they have evidence that students have the knowledge to do so (they scored well on the page of 20 examples using the procedure). In this more traditional type of assessment, students have not been asked to think through a problem situation, decide upon a strategy, apply their mathematical knowledge of concepts and processes, and explain and justify their work as they often are in open-ended questions. The use of open-ended questions also allows students to demonstrate their ability to use mathematics, the second component of the broader definition of assessment. One or two well-crafted, open-ended questions can provide a showcase of students' ability to use and apply their mathematical knowledge and understanding and allow teachers to sidestep the previous scenario.

Another purpose of assessment is to ascertain a student's disposition toward mathematics. This affective component receives little or no attention in more traditional assessment practices. Most elementary classroom teachers can improve learning opportunities considerably if they include affective descriptors in their thinking about their students and their performance in mathematics. The most prevalent of these is the most generic of all—attitude. Also included are persistence, imagination, creativity, confidence, risk taking, motivation, self-reliance, intellectual curiosity, enthusiasm, responsibility, the ability to work with others, share ideas, collaborate, question, and attempt alternative methods. The use of open-ended questions provides opportunities for these descriptors to surface. When a teacher reads the response to the question "What was easy for you in this activity and what was hard?" and learns that it was hard for a student "to share my ideas out loud in the group, but I did it anyway!", information about the child's willingness to take risk has been gained. Teachers are presented with tangible evidence of these somewhat undervalued but very important dimensions of learning. This supports teachers as they work to gain an understanding of the whole student.

Two closing thoughts about the role of assessment are central to the power and the place of using open-ended questions in mathematics. The first is from the NCTM: "Mathematics assessment must reflect what is important for students to learn, rather than what is easy to assess" (1995). It is undeniable that traditional forms of assessment are easier than open-ended questions to construct, administer, and evaluate. The question that must be asked, however, is whether students are being given the opportunity to show what they know and understand about mathematics in ways that are meaningful and relevant. Apropos of that is a saying that was found on the office wall of Albert Einstein: "Not everything that counts can be counted and not everything that can be counted counts." Open-ended questions can help teachers "get at" the things that count so that the mathematical experiences of all students are improved.

SUMMARY

Open-ended questions can help elementary educators attain a number of goals. The use of open-ended questions supports teachers as they answer the calls for curricular reform by engaging students in meaningful mathematical experiences in which they are given opportunities to show what they know, understand, feel, and can apply. The use of open-ended questions can support instructional practice in ways aligned with research-based views of learning and teaching where the focus is on the students who are asked to *do* mathematics and then to explain and justify their doing and thinking. The use of open-ended questions can also help teachers broaden their assessment practice in ways that allow them to gather evidence about the concepts their students "know," the procedures they understand and can apply, and their dispositions toward mathematics.

2

WHAT ARE OPEN-ENDED QUESTIONS?

ASSESSMENT IN GENERAL

DEFINITION OF ASSESSMENT

Before examining the fit of open-ended questions under the overarching umbrella of assessment, it makes sense to think about what assessment is and why we do it. From the many definitions that exist, perhaps the cleanest is the one put forth in the NCTM's *Assessment Standards* stating that assessment is

> the process of gathering evidence about a student's knowledge of, ability to use, and disposition toward mathematics and of making inferences from that evidence for a variety of purposes. (NCTM, 1995)

In that definition, there are three components of mathematics assessment: knowledge, application of that knowledge, and the interest in pursuing knowledge. Knowledge can be thought of as including both mathematical concepts and procedures. The application component can be fleshed out by thinking of it in terms of processes such as problem solving, communication, reasoning, and connections. The third component is one that many may not value as much as the first two which are cognitive in nature. In addition, many find that disposition is much more difficult to assess and are, therefore, uncomfortable addressing it. Disposition can include, among other things, persistence, imagination, flexibility, confidence, willingness to ask questions and to take risks, motivation, self-reliance, enthusiasm, and intellectual curiosity.

It is important to remember that a clear message is sent to students, parents, other teachers, administrators, and the community at large about what is worthy and valued simply by our assessment practice. "What we assess and how we assess it communicates what we value!" (NCTM, 1989). This has real implications for education now, in the future, and even in the past, as is borne out by a thought from Plato, "What is honored in a country is cultivated there."

As we focus on assessment, it is critical that overall mathematical goals and outcomes for students are defined. In general terms, there is no more succinct way to state these goals than was done in the *Standards*, which outlined that all students will learn to

- ♦ Value mathematics
- ♦ Communicate mathematically
- ♦ Reason mathematically

so that they will become:

- ♦ Confident in their ability to do mathematics
- ♦ Mathematical problem solvers (NCTM, 1989)

GOALS OF ASSESSMENT

If those goals are held at the core of mathematics instruction as the outcomes for which we are striving, the next step is to generate assessment strategies that will provide evidence in relation to those goals. As simplistic as that sounds, this is where most traditional assessments fail. For the moment, "traditional assessment" is defined as paper-and-pencil tests that primarily consist of true/false, multiple-choice, fill-in-the-blank, and/or matching items. These tests can certainly offer some information about student learning; however, that information is limited in scope. Further, if this type of test is used as the only or predominant means of assessment, teachers "will be unable to assess many aspects (some would say the most important aspects) of student learning" (Danielson, 1997).

If, indeed, traditional assessments offer little information about student learning and conceptual understanding and even less information about a student's thinking, reasoning, ability to communicate, and disposition, then why are they used at all? One reason is that they can be useful in evaluating certain curriculum goals—those involving low-level skills. Perhaps, however, the major reason is that they are efficient in terms of giving the test, taking the test, and grading the test. The major flaw with this reasoning is that "Mathematics assessment must reflect what is important for students to learn rather than what is easy to assess" (NCTM, 1989). An additional part of this trap is thinking that because you can assess something relatively easily it must mean that it is worthy of being assessed.

"Authentic assessment," "alternative assessment," and "performance assessment" are all terms that exist in contrast with traditional assessment. They are often used interchangeably, although some would argue that each has its own refined focus. For the purposes of this book, however, the terms will be used interchangeably in order to avoid what Tom Guskey, professor of education at the University of Kentucky, refers to as a "tangled thicket of terminology." Suffice it to say that the terms refer to assessment practice that "emphasizes the fact that assessment tasks must authentically represent the way in which the learning has been conducted, be worthwhile, represent the way in which the tasks would be conducted in the world outside of school, and make sense to the students" (Moon & Schulman, 1995).

Given that authentic assessment tasks ask students to do something other than fill in the blank or mark true or false, what are examples of questions that are meant to provide evidence of what students know, understand, and can do with specific information, knowledge, and skills? Carefully constructed open-ended questions can do that and more.

DEFINITION OF OPEN-ENDED QUESTIONS

Most put forth the central idea that an open-ended question is one that "presents students with a description of a problem situation, and it poses a question for students to respond to in writing" (Pandey, 1991). A slight expansion and refinement of that basic premise is that open-ended questions in mathematics are contextualized, interesting problem situations involving one or more mathematical concepts requiring student responses that include at least two of the following components: words, numbers, and pictures or diagrams.

CHARACTERISTICS

It is often easier to gain a sense of an abstract entity by fleshing out its characteristics. A listing of characteristics offers definition and shape to a construct by supplying information about what it is, what it provides, what it requires, what it results in, and what it fosters.

Open-ended questions:

♦ Have *multiple entry points*. This allows for students of all abilities to become engaged in the problem at the level at which they are working. It also allows the teacher to observe and understand student responses along a conceptual and procedural continuum.

♦ Often have *multiple solutions and solution paths*. This validates individual internalization of the problem by students which often leads to unique solutions. These solutions may differ from a "textbook" answer, yet are accurate and perhaps even more valuable due to their uniqueness. This gives teachers an opportunity to see more of what their students are capable of doing.

◆ Require *student decision making*. Control for making decisions is placed within the hands of the students when they are faced with open-ended questions. There are many points to consider, many routes to take, and a variety of ways to explain their work. As a result, teachers gain a sense of how students process information and make decisions.

◆ Generate a *need to communicate*. Students not only need to solve an interesting problem, but are also required to find some way to inform others of some or all of the following: their thinking about the problem, the decisions they made in order to solve it, the strategies they used, and their results. This provides teachers with the type of information about their students that can be used to improve instruction.

◆ Foster *higher-order thinking*. When students are asked to show and explain their work, justify their strategies, make comparisons, and create examples and counterexamples, they are required to move beyond the knowledge level of understanding to the comprehension, application, and analysis levels of understanding, according to Bloom's taxonomy. What a view into students' minds this gives teachers!

◆ Promote, produce, and provide *fodder for pondering*. What a great way to capitalize upon students' natural curiosity! Open-ended questions engender wonder, fostering the "what if" thinking that often leads to original, creative work. This provides teachers with an opportunity to see where their students can go when a multitude of options are open.

THE RANGE OF OPEN-ENDED QUESTIONS

Some open-ended questions can be relatively simple to construct, to set into motion, and even to interpret. Others are more complex in their construction as well as in terms of their implementation and interpretation. The choice is yours and is to be based upon your needs and the needs of your students.

A favorite open-ended question of ours is deceptively simple. It contains four words and allows for a torrent of responses in return: *What do you see?*

This question was first used with kindergartners. A picture was chosen from a book simply because it was rich, beautiful, and interesting. It showed a banquet table at an underwater sea palace. The table was meticulously set, laden with lavish food and drink, and surrounded by many royals and the "common" main character. Because it was set underwater, the picture was brimming with all sorts of fish, shells, sea plants, and so on.

After listening to and enjoying the story, the students were asked to focus on a color copy of the picture on the overhead projector. The simple question *What do you see?* opened the floodgates. Initially, responses were of a generic nature: some people, a king, some fish, fruit, dishes. However, the students soon became both more specific and more quantified: two princesses, nine juice cups, eight big fish, three little pink fish, and so on. As you can see, the question is one with *multiple entry points* in that all children could and did become engaged at their level of operation.

After a whole-class discussion, the students were asked to share their thinking in a way so that when someone looked at their work without having their bodies in front of them, that person would be able to know and understand something about what they knew and understood. The recording sheet simply had a statement of the question at the top followed by directions that students could answer in words, pictures, and/or numbers. Obviously, there were *multiple solutions and solutions paths*. There was no one right answer to this question and no one right way to get there.

As students began to work, they had to consider how they wished to respond to the question. Some used numbers and words, some used numbers and pictures, and some had a combination of all three (see student work samples in Chapter 7). It was evident that *student decision making* came into play in terms of the ways in which they responded and the number of responses, as well as the level of detail included.

The question was interesting enough to the students that they all had a *need to communicate* their thinking. That need was underscored, however, by the requirement placed upon them to share their thinking in a way that someone could know what they knew and understood when not able to ask them in person. They both had to and wanted to find effective and appropriate ways to communicate that thinking.

The question revealed a great deal of what students knew and could communicate. On the simplest level, it was clear who had one-to-one correspondence in place and had the ability to show that consistently. It was also clear who could group and regroup items in a set based on various attributes (four men, five girls, nine people)—evidence of *higher-order thinking* at the Kindergarten level.

Students were engaged by the problem, one that could and did *promote, produce, and provide fodder for pondering*. The students thought carefully, gave answers, thought some more, and gave more answers. They looked for a variety of ways to think, wonder, and reflect upon the question.

SUMMARY

Open-ended questions in mathematics are contextualized, interesting, meaningful problem situations, necessitating student responses that include at least two of the following components: words, numbers, and pictures/diagrams. The defining characteristics of open-ended questions include having multiple entry points and multiple solutions and/or solution paths. They require student decision making and generate a need to communicate their thinking. Further, open-ended questions foster higher-order thinking and promote, produce, and provide fodder for pondering.

3

How to Create Open-Ended Questions

General Techniques

Creating open-ended questions may be even harder than answering them. However, as with most things, the more you do it the better you will become. Another appropriate aphorism is that two heads are better than one, so encourage a colleague to engage in the endeavor with you or to offer feedback on questions you create. Be sure to answer the questions in writing yourself to gauge the difficulty that students will have in framing a response and to capture some of the criteria you will use to evaluate students' work.

Keep in mind that the reason to use open-ended questions is to inspire much more than just fact recall or skills practice. Rather, you are looking for what students know and why they believe it to be valid. As a matter of fact, adding "How do you know your answer is true?" to any closed question is one of the easiest ways to open up student thinking and assessment opportunities. The following are other techniques that can be employed to create open-ended questions.

Jeopardy

As in the game show of the same name, you give students an answer that is appropriate for a question from the topic they are studying and then invite them to create statements or questions that would yield such an answer. This should produce multiple responses and the opportunity to observe how each student is able to apply ideas from recent work.

EXAMPLE

> **Twenty-five cents is the answer. What could the question be?**

Student responses might include:

How much is two dimes and a nickel?

Which coin is greater than a dime?

13¢ and 12¢ = ?

How much money do you have?

EXAMPLE

> **The probability is 1/4. What could the question be?**

Student responses might include:

What is the probability of spinning blue on a color wheel divided into four equal sections with one colored blue?

What is the probability of getting two heads when tossing two coins?

What is the probability of picking a month that begins with a J?

Such solutions have many possible questions. If students are encouraged to give multiple responses and to think expansively, much can be gleaned about the range of their thinking and their ability to make connections between ideas.

RESOLVING A DISAGREEMENT AND CORRECTING MISTAKES

You might take the fodder for these types of questions directly from students' papers. Choose the erroneous work of one or two students as a point for discussion. Be sure to change the names or ask the authors' permission first.

EXAMPLE

> **John believes the answer to 24 + 37 is 51, but Katrina is almost positive it must be 511. Who is correct? Explain your thinking in writing.**

Although offering two incorrect solutions may seem like a trick question to the students, it is a particularly effective method for finding other students who have misconceptions as they try to defend one of the positions. It is also helpful for identifying students who are able to explain to their peers why both solutions are incorrect.

A classic geometric misconception is used next in a similar fashion.

EXAMPLE

> **Maria tells Toni she has five rectangles in her pattern block design. Toni says she is wrong because she sees three rectangles and two squares. What will you say to resolve the disagreement?**

Along similar lines, one teacher asked a class of fourth graders to identify all the different types of mistakes one could make in a specific two-digit by two-digit multiplication problem. A long list was created by combining the ideas from all the students. Besides being a great open-ended question, it helped students learn to reflect on their own errors. Often students have difficulty identifying their own mistakes. It is important to help them develop explicit techniques for checking their solutions for reasonableness and for error analysis.

FROM THE REAL WORLD

We always tell students that mathematics is everywhere, but do we create an environment that validates this perspective? Look for things in your home or the school that will entice students to find many different solutions as well as real-world applications. Menus, sales flyers, or magazines are good sources.

EXAMPLE

> ♦ **If school lunches came from this restaurant, what might they be and how much would they cost?**
>
> ♦ **What are the possibilities for spending between $5 and $10 at this store?**
>
> ♦ **What can you say about the relationship of space devoted to articles and to advertisements in magazines?**

You may want to encourage students to bring in props or make drawings for their reports on a number, measure, or shape treasure hunt such as the following:

EXAMPLE

> ◆ **Where can you see mathematics at work on your way home?**
>
> ◆ **What shapes are in the room, and how are they used?**
>
> ◆ **What units of measure are in your cupboards at home, and why were they selected for those particular products?**
>
> ◆ **What mathematics do others in your family do during their day? Is it like any work you have done?**

These types of questions can be used to evoke interest in new mathematical ideas as well as to assess what the students have already learned. For example, if individual students did not find all the different measuring units identified by other students, they are likely to want to be introduced to those they may have never noticed.

TELL ME ALL

One of the teachers we worked with began her open-ended question odyssey by simply asking students to write about what they knew relating to a topic before and after a unit of instruction. She related that it was slow going at first, but by the end of the year, the students were quite proficient, usually filling more than one page. She also commented that with this approach, she knew more about her students' abilities than she ever thought possible.

With questions this open, it is often effective to give or elicit vocabulary words that could be included in the response. Graphic organizers might also be used to help students capture all their ideas well.

EXAMPLE

> **What things do you know about fractions?**

Fifth graders tried this question. The words and phrases they brainstormed initially only included: thirds, fourths, numerator, denominator, part of, whole, shading. Their first attempt at writing about fractions highlighted their misconceptions and confusion but was helpful in planning follow-up experiences. By the end of the unit, the students' responses were much richer, and they were pleased to see their own growth in knowledge.

Younger children can do similar work by offering pictures as well as words. These can be used by the class in writing a story about the mathematics they explored during a week.

EXAMPLE

> One teacher read from and made available several books about shapes so that students had models. She engaged the students in several hands-on experiences with pattern blocks. As a nice connection among literature, art, and geometry, she had pairs of students work together to create a story.

With some students it might be necessary to begin to learn how to respond to a tell-all question by asking for stories about a single day.

EXAMPLE

> Write and illustrate a story involving pattern blocks.

The tell-all type of question helps students learn to reflect on and synthesize the mathematical ideas that they have been experiencing. If students are to become lifelong, independent learners of mathematics, these skills of reflection and synthesis should be a major focus in the schooling experience. Work with them should begin at a young age and be fostered each year.

ADAPTATIONS FROM TEXT QUESTIONS

CHANGE THE WORDING

If the goal of using open-ended and open-response questions is to examine students' thinking, then we must choose words that inspire thinking. Mathematics textbook problems are usually prefaced with words that only suggest performing a learned routine: add, multiply, solve, and so on. Change the text experience with words you may have associated with thinking skills in other disciplines.

EXAMPLE

- With "3 + 4 = ?", ask students to *draw* a picture that proves the answer is 7.

- With "round 23 to the nearest 10," have them *explain* when it might be advantageous to use 20 and when rounding to 30 might be more advantageous.

- With a page of division problems, ask students to *predict* which will have quotients greater than 10 and to *explain* why they chosen particular examples.

- With a whole page of varied problems, ask students to explain how some problems are *alike and some are different*.

- With word problems, have students *restate the problem* differently.

- With number sentences, have students make them meaningful by *creating word problems* that might spawn such number sentences.

- With any problem, ask students to *generate other problems* that would yield the same answer.

Sometimes you will find ideas for open-ended questions embedded in the teacher hints or dialogues on the side of the lesson pages. Rather than telling what is suggested, pose a question that will have a similar effect but will be more thought provoking.

EXAMPLE

- Text: "Remind students that multiplication is repeated addition."

  - Question: In what ways are addition and multiplication alike?

- Text: "Tell students that 44 means 4 tens and 4 ones."

  - Question: How are the fours in 44 different?

From Illustrations

A picture may not elicit a thousand words from a young student, but it may be worth more than a typical textbook question. Look for an illustration and consider how it could relate to questions other than the one being asked. Here are some questions that help students to see a collection of squares from many different perspectives and to make connections across several different mathematical domains.

Example

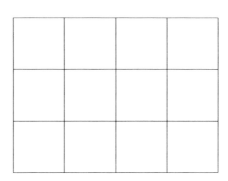

- ◆ Show all the different rectangles you can create by folding along one of the line segments of this rectangle.

- ◆ Explain to someone how to make this drawing.

- ◆ Draw pictures to show the ways can you shade one-third of this figure.

- ◆ This is one of the shapes in a pattern of growing grids. What might come before and after it? Describe the pattern.

- ◆ What other ways could these tiles be arranged to make a rectangle? How does this change the area and perimeter?

Most commercial mathematics programs do require adaptations to make their questions more open-ended. To insure that there is sufficient instructional time for all the areas of mathematics offered by commercial texts and open-ended adaptations, consider reducing the amount of time your students spend working with "naked numbers;" that is, exercises that come without a context and yield only one right answer. Students need meaningful experiences to make sense of the mathematics they do. It is important that we help them understand "when and how they will use this stuff."

FROM GOAL STATEMENTS AND STANDARDS

Writing open-ended questions from goals can seem difficult because goal statements are more abstract than day-to-day experiences. Perhaps this is why the NCTM created *Addenda Books* (NCTM, 1992) . You will find many examples in the seven-grade level and four topic-specific books in this series that relate directly to the *Curriculum and Evaluation Standards* (NCTM, 1989).

Many states have produced documents relating to their statewide assessment programs that include samples of open-ended questions. You can often download such documents by connecting to the state education office through the Internet. Information about some of these documents is contained in Chapter 6, "Where Open-Ended Questions Are Found."

To create open-ended questions from your district's goals, start from a broad goal statement rather from each individual behavior or skill. Next, consider a few behaviors that you think will show students' grasp of the idea(s) in the goal statement. Then think of an open-response task that would afford you the opportunity of assessing how students used the behaviors to find a solution.

EXAMPLE

♦ **Goal**

- **Students will exhibit understanding of the place value system.**

♦ **Behaviors**

- **Students can represent the digit values.**

- **Students can name the places up to a million.**

- **Students can order numbers.**

- **Students can identify real-world uses for numbers.**

♦ **Task**

- **Look through the newspaper during the next week. Find three examples of dollar amounts that are greater than $1,000. List them in order from smallest to largest. Explain what each amount would look like as a collection of bills. Also include an explanation of why the amount was cited in the newspaper;**

OR

- **If you have one of every coin, what are four different things you could afford? Order the amounts from most to least expensive. Draw pictures and show the amounts in coins.**

Both are fairly simple tasks that allow children to exhibit their knowledge of numbers in the world as well as several numeration concepts. As you can see from the example, one of the bounties of using open-ended questions is assessing several behaviors with only one question.

STUDENT-CREATED QUESTIONS

Much can be learned about students' knowledge by the types of questions they choose to offer. Some teachers review for tests by having students create questions that they themselves think should be asked. Students typically offer closed questions but may be able to create more open versions if they have had opportunities to respond to different types of questions. Students may be inspired by using techniques similar to those shown in this chapter. Students can be given a solution and a goal of the unit, say "18 inches" and "understanding subtraction."

EXAMPLE

Use the "jeopardy" model and create a question that would have the answer 18 inches and that must require some thinking about subtraction.

SAMPLE SOLUTIONS

A sample closed question is, "What is the difference between 36 inches and 18 inches?" A related open question would be, "Name two things in the room that are different in length by about 18 inches." Though both questions show a good understanding of the concept of difference, the capability to find the correct answer, and the ability to reason backwards from a solution, the more open question addresses several of the standards in *Assessment Standards for School Mathematics* (NCTM, 1995, p. 18) :

3. Mathematics as Reasoning
5. Estimation
7. Concepts of Whole Number Operations
8. Whole Number Computation
10. Measurement

Research indicates that students who create problems perform better on tests of problem solving than other students who have not had the experience. Even those who have had explicit training in techniques of problem solving do not perform as well as students who have problem-creating experiences. Also, students enjoy solving problems created by their peers. Consider using student-created questions on homework and tests.

QUALITY CONTROL

Sometimes it is hard to know whether a question you have constructed will be a valuable assessment tool. Many of the sample questions in this chapter are stated very simply but yield a wealth of responses if students are encouraged to find many possibilities. You could argue that both of the following questions are open, but the latter will expose more of the students' thinking.

EXAMPLE

♦ **Name some of the even numbers.**

versus

♦ **What are some different ways to find all the even numbers on the hundred chart and the number line?**

Collison (1992) offers the following questions to help you evaluate whether you have asked a question that is likely to yield rich responses. Does the question:

♦ Focus on essential curriculum concepts?

♦ Lead to other questions?

♦ Tap real-world situations (if possible)?

♦ Allow students to work together to pose solutions?

♦ Allow for multiple pathways to a solution or multiple solutions?

You will probably get more robust answers to questions if they are intriguing to students. Be sure a question offers a new perspective or challenge, especially if it relates to work that has been ongoing in the classroom or was done last year. As an alternative, consider asking the question before students have learned the appropriate skills. This can stimulate their interest in the topic and provide you the opportunity to observe their problem-solving skills.

EXAMPLE

♦ **About how many breaths have you taken in your life?**

♦ **How long is the hall?**

These questions could be used to encourage students to learn about and use different units of measure while they are practicing measuring and computing skills. The large numbers the investigations produce will provide a challenge and an incentive to learn about conversion between units.

If students are confused and unable to get started with a question, it can be a message to you that they are not quite sure what is being asked. The difficulty may not rest with the quality of the question but in the students' understanding of the parameters of the response that will be acceptable. Questions of only a few words may be easy for all students to read but may warrant coaching. Here are a few tips:

♦ Start students with a title sentence or even a list of words to include in their response.

♦ Check to be sure students have an understanding of the content of the question. Ask them to restate the task in their own words.

♦ If the task is very open, one or two examples may be needed to help students frame their responses.

EXAMPLE

♦ **What are different ways to show 20?**

♦ **What are different ways to show ¼?**

SAMPLE SOLUTIONS

20 can be made by:
1 + 19
2 BaseTen rods

¼ on the calculator looks like .25
¼ is 2 shaded squares out of 8

♦ Suggest where to find important information. For example, "You may use calculators and blocks to help you think about possibilities."

♦ Give a sense of the scope of the expected response.

"There are many possible answers. Work by yourself to think of a few and then make a list of everything your group has thought of together."

♦ Post a rubric of what you value in a response.

For more tips about how to help students become comfortable with open-ended questions, read Chapter 4, "How to Establish and Support an Open-Ended Question Environment."

SUMMARY

Like many things, creating open-ended questions improves with experience. The first few attempts may be a struggle for both the teacher and the students, but persistence will be rewarded with significant information about students' knowledge and process skills. Creating open-ended questions is easily accomplished by changing the wording of closed questions in texts or asking students to relate a closed question to a real-world situation. Having students describe errors in a piece of work or explain a process for finding the correct answer also produces multiple solutions.

When we do real-world mathematics, we must often use several different skills, including reflection about and synthesis of things done previously. Open-ended questions that can be used to assess several things at once are time efficient and better reflect why we learn about mathematics. A few open-ended questions that incorporate most of the skills in a unit can inform the teacher about student progress and motivate the students to investigate new ideas. Such questions can be as simple as "tell me what you know about _____."

4

How to Establish and Support an Open-Ended Question Environment

At School

Did anyone get the same answer in a different way?
What is the same or different about your two ways of doing this?
Did anyone get a different answer?
How did you get your answer?
What do you think helped you get your answer?
Tell me what you are thinking.
What would happen if…?
Is there a pattern? What is it? or Why not?

adapted from Rowan and Robles (1998)

It is likely you have already used questions similar to those just described. They are clearly very open-ended questions that will yield a variety of responses and show students that you value hearing their ideas. If you have asked such questions, you are well on your way to creating an open-ended question environment. The goal should be to establish a classroom climate where differing views are shared, valued, and expected at all times. It is important to give students implicit and explicit messages that support this goal.

27

DIFFERING VIEWS:
ORAL RESPONSES

All children learn the rules of a game quickly. In mathematics class, the game is often to ascertain exactly what answer the teacher wants and to be the first one to say it. One of the easiest ways to change that notion of mathematics learning is to expect students to defend the validity of their responses, or, in kidspeak, "prove your answer is right by explaining what you did."

Even if there is only one correct solution to a problem, there is usually more than one way to arrive at that solution. Many teachers require a two-part response to most questions; one for the correct answer—the "what"—and one for a reasonable explanation—the "why." The following classroom vignette helps to illustrate the potential richness of such an approach. Though these responses do reflect actual third graders' thinking, they were edited to make them easier to follow.

EXAMPLE

65 + 57 = ?
Why do you think your answer is right?

SAMPLE SOLUTIONS

- "I think it equals 112 because 5 plus 7 is 12 and 6 + 5 is 11."
- "I think it equals 122 because I put the bigger number above the smaller, added the 5 and 7 and got 2 in the ones, carried a 10. Then I added the 6 and the 5 for 11 with the carried 1, which makes 12 tens or 1 hundred and 2 tens and 2 ones."
- "First I ballparked the answer by thinking of 50 + 50, so I knew it was more than 100. Then I counted by 5's from 65 to 100. It took seven 5's, or 35 from the 57, leaving 22 more beyond the 100. So I think it is 122."

When the teacher entertains all three as possible correct responses, each child's right to participate in the game of creating mathematics is validated. As a result, the game becomes more interesting than just finding out what the teacher wants. Wouldn't you say the last answer is as valid as the second, and perhaps more elegant, especially if a person is trying to do the work in his or her head at the supermarket?

Having students share responses can help all the students learn about self-editing. In the class when this work was done, by the time the third solution was offered, the first child had already independently amended her thinking as to why her solution was not correct. It also appeared that many of the remaining students in this class were evaluating their own thinking against these three presentations.

A rich open-ended question environment can be created simply by expecting and accepting more than one way to arrive at the solution. The only effort it requires is allowing children to offer their ideas before any one solution is validated as reasonable.

Contrast that scenario with what happens in a less accepting environment. Unless students know they must give their thinking with their solution, they often offer the first thing that pops into their head. If it happens to be correct, many teachers accept it and move on to the next question without waiting for the other students to think through a variety of viable scenarios. For some students this does not even leave them time to rethink the wrong response they would have given if called upon by the teacher. On the other hand, if the first impulsive response to a question is wrong, teachers often just continue selecting students until someone does offer a correct answer. This reinforces the idea that the game is won by the student who figures out what the teacher wants first, rather than by the student who has thought deeply about the problem. Before moving on to another student, consider asking an impulsive responder to explain how he or she arrived at the solution.

The first scenario is one that will encourage students to be persistent and creative in finding solutions to open-ended questions. The downside about using such a scenario is the amount of time it takes to process students' work orally. The upside is that less time is needed later for rote practice because the quality of the work and the retention of the important ideas are more readily attained when students have the opportunity to work through their own thinking.

DIFFERING VIEWS: WRITTEN RESPONSES

Going over work orally can be very time consuming, may spotlight only selected students, and may not give you the full picture of each child's knowledge. For this reason, it makes sense to ask students to offer the what and why of their solutions in writing. This work is likely to be more valuable for your students and you than workbook pages that offer only practice of the same skill over and over again. All students have something they can show on paper, and all can become more skillful in conveying their thinking through this medium.

Start small. Anything that yields more than just a numerical answer is often a worthwhile beginning. Be sure to leave yourself enough time for coaching students about how to improve their responses rather than asking quantities of open-ended questions that make the experience feel burdensome. One or two open questions a week is a good beginning if you are requiring thorough written responses.

Written responses give the message that you expect work from all students, but words alone may not tell the whole story about what individual children know. Research suggests that there are several different learning styles present in every classroom. Not only does this mean that students should experience a variety of intake options, such as auditory and visual presentations, but output options as well. From time to time, offer students a choice of reporting mediums: posters, video- or audiotaped presentations, dramatizations, or even poems.

Example

> **Write everything you know about any interesting number at the end of a unit on number theory.**

Fifth graders who tried this question developed "guess the number" clues on well-decorated posters, some wrote creative stories personifying the number, and one used a talk-show format with another student as an interviewer. One of the most unusual was a bulleted list showing all the characteristics the number 13 didn't have: not even, not divisible by anything except itself and 1, not usually the area of a rectangle, and so on.

Deep Thinking

The amount of time needed to respond effectively to open-ended questions can vary from student to student and from problem to problem. Even those students who are typically fast with a response can benefit from extended periods for reflection about a problem and from solving more than one problem of a similar type. One approach is to give time for a question on one day and suggest that everyone continues to work on his or her response for homework. However, if using the exercise to evaluate each student's progress, you might rather give time the next day for more work on the problem so parents aren't involved in the response.

"As teachers of writing have known for years, reflection on written communication is greatly enhanced when students have the opportunity to revise their work" (Wilson, 1995). Wilson suggests a two-stage test system used in the Netherlands where students revise their first draft after receiving teacher feedback. She believes this encourages students to think at deeper levels. It's not always possible to offer revision tips to each student but most students can do it for each other if the criteria are clearly stated. Modeling this idea with one or two volunteers' papers is an effective way to introduce the idea of peer editing.

Time for revision gives students the message it takes careful thought and persistence to produce good mathematics. Some teachers allow students the opportunity to earn more points or a higher grade by reworking their papers at least once. Such techniques send the message it is their best work not just a solution that is valued. The goal should be to help all students reach the highest level of achievement even if it takes several attempts.

The depth of thinking and retention are also aided by giving students a second very similar problem. Aren't we all improved through multiple experiences, and isn't everyone capable of producing good work with persistence? It helps to be explicit about these ideas when giving students the second, and sometimes third, related question.

EXAMPLE

> ♦ **What things come in tens?**
>
> ♦ **What things come in hundreds?**
>
> ♦ **What things come in thousands?**
>
> OR
>
> ♦ **How are these two shapes alike, and how are they different?**
>
>
>
> ♦ **How are triangles and quadrilaterals alike and different?**
>
> ♦ **How are polygons and polyhedra alike and different?**

When doing related problems, students should be explicitly reminded to review their work from the earlier questions before creating a response to subsequent versions. This will generally engender a more robust response. If it does, be sure to comment on the change to encourage the concept of reflecting on past experiences.

RESPONSIBILITY FOR LEARNING

People perform better when they know the goals, see models of good work, and know how their performance compares to the standard. One of the best ways to foster student independence is to be clear about what successful work looks like before students finish responding to an open-ended question. If students themselves help to set the criteria, the standards are often clearer to them because they contain their own language.

Second and third graders can be quite adept at identifying assessment criteria (a rubric) for their work. At first, it helps to have students spend some time writing a response before having a discussion of what important things should be included and how they will be credited. Sometimes the teacher needs to help with the specific mathematical ideas, but students usually recognize the need for coherence and correctness in a response. Be sure to allow students an opportunity to rework their responses once the criteria have been specified.

EXAMPLE

> **Explain how to solve the problem $.46 and $.75 with pennies, dimes, and dollars.**

A grade three class that did the above example created this rubric:

> 2 points for the correct solution
>
> 2 points for a clear, well organized explanation of how to do it
>
> 2 points for talking about what carrying/trading means

One student offered the idea about carrying. It was something that had been emphasized that week and something which should have added to the list had even if no one offered it. Usually the final criteria list results after some synthesizing and negotiation. It is more effective, especially with younger students, to keep the list short. It is helpful to take the criteria setting a step further by asking for, or giving examples of, things that would provide evidence of meeting the criteria. For example, a diagram of the pennies and dimes to aid the explanation or information about why a 1 was placed above some of the columns might be offered for the example just given.

Students taking charge of their own learning also means seeking resources from models, books, past work, and peers rather than relying on the teacher. Working with others sometimes creates too much reliance rather than personal perseverance. Sometimes students don't understand how to use such resources fairly. The THINK, PAIR, THINK model works well by helping students learn to take personal responsibility while expanding their ideas by using those of others. It also allows both internal thinkers (those who need quiet) and external thinkers (those who need to verbalize) a venue for doing good work.

> THINK—Take 10 minutes to organize and begin to write about your own ideas.

PAIR—Explain your thinking to your partner then listen carefully to his/hers.

THINK—Take 10 more minutes to think quietly about how you can expand your ideas by using some of your partner's thinking. Be sure to give your partner credit by explaining how his/her ideas helped you expand upon your own when making or writing your final report.

Try this approach with one or two of the following examples:

EXAMPLE

Verify your thinking with pictures and words.
- How many jelly beans are in this jar?
- How many squares cover your desk?
- How many milk cartons could fill the room?

WORKING COLLABORATIVELY

Problems that have many solutions can be daunting for some students to focus on long enough to see a pattern. Consider using cooperative group techniques to empower students to investigate questions that might otherwise cause frustration for some. You might want to add a SHARE step to the THINK, PAIR, THINK model with the examples already suggested or those given next.

EXAMPLE

- What are ways you can make 100 on the calculator when the 0 key is broken?
- Which fractions equal .25 when put on the calculator?
- Which fractions are less than ½?

Because these are truly open-ended questions, with unlimited answers, you may want to put up a large poster-size paper that can be added to over the course of a week. This helps to insure that there is room and time for all to contribute to the product.

Many suggestions for working in cooperative groups are well suited to working with open-ended questions:

♦ Role assignments, such as recorder, reporter, supervisor, and investigator

♦ Arbitrary selection of the group reporter at the task's completion so that everyone takes responsibility for understanding the solution process

♦ "Don't ask me till you have queried everyone in your group" type responses to student questions

To help students appreciate the benefits of working collaboratively, have them answer the first question alone, but then experience the expansiveness of group answers with the second and third questions in an activity like the following.

EXAMPLE

1. **What are all the amounts of money you can make with 2 coins?**

2. **What are all the amounts of money you can make with 3 coins?**

3. **What are all the amounts of money you can make with 4 coins?**

AT HOME

Few American adults have had the opportunity of exploring open-ended questions through their own classroom experiences. It is not surprising, then, that parents are likely to be confused and anxious about their children being given such questions for homework. Because research indicates that parents' attitudes about mathematics can affect their children's success, it is important to help parents understand the merits of asking students to think deeply about challenging problems.

Though it is possible to reserve open-ended questions for the classroom, parents would miss seeing firsthand information about the potential and excitement engendered when children create mathematics. Further, observing a child's processing of problems can provide more feedback about his or her mathematical strengths than a letter grade or comment from the teacher. We need to make parents allies in changing the way mathematics is taught in schools so they can help to interpret changes for the broader community that is responsible for education funding. For all these reasons, consider the following ideas for communicating with parents about open-ended questions.

ENLISTING SUPPORT

Workers in most fields, including mathematics, are often called upon to validate their reasoning about an approach they are using and/or to adapt to an approach that is new to them. Children need to acquire flexibility in their approaches to problems. We all need children to learn to be inventors of solutions to complex problems that they will face in the future. It is never too early to encourage this capacity.

Machines now perform "shopkeeper" arithmetic skills with greater efficiency than most humans. Studies show that up to 90 percent of instructional time for mathematics in elementary schools is spent on learning and practicing these same skills. Children still need to know facts, be able to ascertain a reasonable estimate for a computational procedure, and understand the procedures that yield correct answers. However, the limited instructional time for mathematics learning available in a school year must be better utilized so that all domains of the subject are addressed well. For instance, the information-age requires that citizens be able to interpret and represent data fairly and accurately. Much more time must be given to the study of statistics and probability during school years if we are to have an informed, judicious electorate in the future.

Mathematics classrooms devoted to replication of routinized skills are yielding students who perform at or below average even on the computational parts of international tests. "Even when correctly learned, purely procedural knowledge—the ability to implement mathematical algorithms without underlying conceptual knowledge—can be extremely fragile" (National Research Council, 1989).

In a study done during the *Third International Mathematics and Science Study (TIMSS)* (Stigler & Hiebert, 1997), where Japanese students scored at the highest levels in all mathematical areas, it was noted that Japanese parents believe it is attitude and persistence, not aptitude or teaching that determines their child's success. They also found that Japanese teachers have a lesson goal of improved thinking 70 percent of the time versus only 22 percent in the United States. Our students need to believe that becoming lifelong learners of mathematical ideas requires hard, independent work. They must become accustomed to explaining their thinking when doing mathematics so teachers and parents can assess whether intervention is warranted.

For all these reasons, it is important that children grapple independently with challenging open-ended meaningful problems. We need to help parents learn to accept that mistakes and struggle are part of learning new things.

HELPFUL TECHNIQUES

Even when parents are convinced that open-ended questions are a path toward achieving many of the goals described in this chapter, they will still want to help their child when their child is struggling. It is wise to offer them suggestions about how to do this. Parents may only know the techniques employed by their own teachers or parents. Here are some suggestions you may want to put in a note home:

◆ Applaud your child's struggle and persistence, explaining that these are attributes that show they are good mathematics students.

◆ When asked for help, inquire about how the child perceives the task rather than interpreting it for your child. Resist showing your child the way you learned. Help your child see errors in his or her thinking by asking questions. If your child appears to be confused or heading in the wrong direction, try to help your child find another approach that will make sense to him or her. Many different procedures are usually acceptable.

◆ Model use of references and models. Look things up in past homework exercises. Obtain a mathematics reference book that you can use together when your child is unsure of information. Offer your child counters, rulers, coins, and the like for modeling the situation, and also suggest drawing pictures and diagrams.

◆ Encourage your children to write a note about where he or she is stuck if none of the above tips help. It is likely the next day's class will clarify the issues, but it is important for the teacher to understand the child's difficulty so the intervention is appropriate.

◆ Help your child understand that mathematics is a tool used to help organize and understand the world around us. Like explorers, we use it to investigate new ideas or to validate known ones. We often begin with few road maps, improving upon the course we choose with each related experience.

SUMMARY

Opening up a classroom environment to encompass different perspectives can be done simply by taking all responses to any question without acknowledging a correct solution and allowing students themselves to argue their cases. Asking several students to explain why they believe a response is correct illustrates that even a single solution may have multiple solution paths. Written responses about why a solution is correct can allow every child to participate and provide the teacher with information about what intervention in the learning process is needed. Asking students to revise their first responses helps them to become responsible for their own learning. Clearly specified criteria make this personal editing work more meaningful and effective.

Multiple experiences with similar problems helps students produce robust responses to questions with multiple solutions. Robust solutions are also facilitated by listening to and incorporating the thinking of others.

Parents can help students open up their thinking process but they will probably need justification for this new approach to mathematics learning. Specific strategies for aiding students with open-ended questions will also help parents become allies in preparing students for the challenges of the new century.

5

HOW TO EVALUATE
OPEN-ENDED
QUESTION
RESPONSES

PURPOSES OF ASSESSMENT

Open-ended questions provide a great deal of information for a variety of purposes. As a result, evaluating responses to open-ended questions simply in terms of "right" or "wrong" would be shortchanging their potential. This is especially true given that an underlying premise of assessment practice is to create opportunities to offer feedback for reflection and revision, serving three of the purposes of assessment from the National Council of Teachers of Mathematics (NCTM, 1991):

- ♦ Monitor students' progress

- ♦ Make instructional decisions

- ♦ Evaluate students' achievement

ASSESSING OPEN-ENDED
QUESTIONS IN GENERAL

First, a disclaimer is necessary: Not every open-ended question needs to be assessed and evaluated formally. It is situation-dependent. An open-ended question that functions as a pretest can sometimes be used to make instructional decisions (e.g., "Can you tell, show, or write what you know about multiplication?"). The purpose of this pretest is to help determine the relative areas of strength and understanding of concepts. There may be no need to assign a formal label or number to each student's response in this situation.

There are times when anecdotal observations are enough for a teacher's purposes. For a specific open-ended question, for example, a teacher may be looking to see if students have generated more than one way to solve the given problem. With sticky notes in hand (or index cards, clipboards, mailing labels, etc.), the teacher may roam the room observing and noting whether students have found more than one solution path.

There may be times when the teacher's purpose is to ascertain the level of response students will give to the question "How did you figure that out?" This can be a 30-second interview and can also happen somewhat informally as the teacher walks through the room speaking to students on an individual basis.

There are, of course, times when a more formal assessment of open-ended questions is needed. This can be done either holistically or analytically and can provide important information about what students know, understand, and can apply in terms of mathematical concepts, processes, skills, and dispositions.

HOLISTIC ASSESSMENT OF OPEN-ENDED QUESTIONS

In broad terms, holistic assessment of open-ended questions provides an overall feel for the student's responses—that is, how well the students fulfilled the general purpose(s) of the question. Holistic assessment provides more of a *gestalt* in terms of giving the whole picture.

A movie that was great, just okay, or awful is an example of a holistic rating in a noneducational context. The individual criteria of movies have been combined or "lumped" together, so to speak, to give a general "feel" for the movie. A student response to the open-ended question, "How would you explain subtracting with two-digit numbers to a younger student?", could be assessed holistically, for example, by determining if the response in overall terms is "exemplary," "satisfactory," or "incomplete."

While this is sometimes what is called for in certain educational situations, more often than not, holistic scoring collapses much of the pertinent information so as not to provide sufficient detail to students about the relative strengths and weaknesses of their work. So, while it may serve the purpose of evaluating students' achievement at that moment in time, holistic assessment does not always offer specifics about student progress or about what can be done to improve the work.

ANALYTIC ASSESSMENT OF OPEN-ENDED QUESTIONS

Assessment of open-ended questions can be analytical with points either awarded or subtracted for individual elements of the student response. An analytic assessment clearly defines the various criteria and the corresponding levels of performance. The acting was terrific in the movie, but the pacing was uneven, and the plot was totally unbelievable. An assessment such as this paints a more complete picture of the movie.

The same holds true in the classroom. The student response to the subtraction question can be assessed analytically by evaluating various criteria. For example, the response can be evaluated in terms of the clarity of the explanation, mathematical concepts and vocabulary, or examples and counterexamples. This type of assessment can provide specific information to students, teachers, and parents about what the student has accomplished and to what degree.

DEFINITION OF RUBRIC

There are many definitions of a rubric. According to *The American Heritage College Dictionary*, a *rubric* is "a short commentary or explanation of a broad subject." That basic definition has been shaped in educational practice to stand for a set of guidelines used to give a score of some kind and offer a commentary or explanation of the work.

The point needs to be made that a rubric as a set of criteria used to judge, evaluate, critique, and assess student work can be used either holistically or analytically in assessing open-ended questions. Once the criteria have been established and the essential mathematics has been identified, the evaluator can look at the criteria as a whole and give one score for the response, assign a score to each of the criteria sets and let multiple scores be given, or combine the scores from the criteria sets to arrive at one score for the question—an "average," so to speak.

USING RUBRICS TO ASSESS
OPEN-ENDED QUESTIONS

With a rubric in place, the evaluator has several options. Student responses can be evaluated in ways that best meet the needs of the situation. Given the criteria, student responses can be evaluated in terms of various levels of performance.

TWO-WAY SORT

Some open-ended questions can be assessed on a semiformal basis. There are times when a simple sorting of student responses can be accomplished on a two-point scale. This is often the case when teachers are looking for information that will inform their practice and help to monitor student progress. Responses can be separated into two piles: acceptable or need work (traditionally known as pass/fail but could also be thought of as complete/incomplete). This way of evaluating a question can provide some useful information depending upon the purpose of the task. For example, an open-ended question focusing on students' ability to be self-reflective could ask students if a task was easy or hard for them and why. Student responses may or may not be complete enough to provide enough critical information and, therefore, "need work," be "incomplete," or "need help."

THREE-POINT TARGET

This type of sorting allows for assessment that is more refined and defined. Again, it can be used either holistically or analytically as the evaluator thinks of a target such as the one used in archery. A response to an open-ended question can be evaluated on the basis of the established criteria and then determined to be a "bull's eye," on the outer rings, or not even on the target. Instructionally speaking, this could translate to three categories:

- ◆ Exemplary: complete understanding, no revision needed
- ◆ Satisfactory: some understanding, some work needed
- ◆ Incomplete: little or no understanding, needs help

FOUR AND MORE

Rubrics with four or more levels of performance can offer finer distinctions between responses and, thereby, provide more specific information than rubrics with three or fewer performance levels. They can be used to monitor student progress in greater detail and can highlight elements of student achievement in ways that give constructive and very specific feedback to students.

A drawback to using rubrics with four or more performance levels, however, is that they can sometimes be unwieldy and time-consuming. Because the distinctions between levels are so specific and well refined, it can take a great deal of time to determine the exact level of performance for each of the criteria.

RUBRIC CONSTRUCTION

Rubrics can be constructed by individual teachers, a team of teachers, teachers with students, or students working individually or with a group. There are advantages to each of these scenarios.

INDIVIDUAL TEACHERS

When a teacher is working individually to develop a rubric for an open-ended question, he/she is forced to think deeply about what kind of information can be gained after looking at student responses to the question. Identifying the mathematical expectations of the task, identifying what else is important about the task, thinking about how many levels of performance there might be and what constitutes a specific performance level all play into creating a rubric. This is an exercise that helps teachers reflect upon curriculum, instructional and assessment practice, and, quite often, leads to insight and growth for both teachers and students.

TEACHERS WITH TEACHERS

There is quite a bit to support how beneficial it is for teachers to work together to construct a rubric for a specific open-ended question or task. What is almost immediately evident is how differently teachers approach such a task. Identifying the criteria that are most important in terms of a specific question can open the floodgates of discussion. Then, determining levels of performance can add another dimension to the discussion. There is no doubt, however, that although it is more than likely that teachers will disagree about many points, the discussion that evolves is an engaging and enriching one. The bottom line is that teachers can and do learn from each other, and the power of collective expertise cannot be denied.

STUDENTS

Students can be involved with rubrics in several ways. They can participate in writing the rubric under the direction of the teacher. This can be done as part of a whole-class discussion in which the criteria sets are identified along with corresponding levels of performance. This should be done before the students complete the open-ended question so that each student is aware of how a response will be evaluated.

A point needs to be made here about the seemingly covert operation of assessment. Traditionally, students do not know until after the fact what they need to do to "get a good grade." This information is not at their disposal—to their disadvantage. When students are aware of the criteria that have been set, they are more likely to meet them; if not on their first attempt, then, certainly, with a rubric in hand, they can and do come closer on their next try. Assessment practice need not be a secret!

Depending upon the age level of the students, there is also value in students working in small groups to develop a rubric for an open-ended question. This has many positive effects including identification on the parts of the students of important mathematical concepts, processes, and skills.

POINTS TO REMEMBER ABOUT RUBRICS

   ◆ Number of criteria

   There is no set number of criteria for each and every open-ended question. Generally speaking, however, it becomes far too cumbersome if more than six criteria are used. In most elementary situations, four criteria can usually define a question.

♦ Categories of criteria

Each of the criteria sets can represent several elements. For example, to continue with the movie analogy, when analyzing the acting, the various components could include the female lead, male lead, supporting female, supporting male, etc. It is important that these sub-categories flesh out the set and reflect subsets that are significant.

♦ Number of levels of performance

Rubrics can have as few as two levels of performance to as many as ten. Rubrics are both situation-and teacher-dependent. If a teacher's purpose is to provide a detailed level of refinement between categories of performance, then the greater the number of levels. For most purposes in elementary classrooms, the usual number is three to four.

♦ Names/titles/descriptors for levels of performance

This is a matter of personal preference. Some practitioners like to use numbers (3, 2, 1), with the highest number on the scale indicating the highest level of performance, while others who use numbers assign the number 1 to the highest quality. Other scales use terms such as expert, practitioner, and novice; exploring, developing, and achieving mastery; and so on.

FRAMING A RUBRIC

THE NAKED FRAME

For our purposes, we have used a rubric for the sample open-ended questions presented in this book with four sets of criteria and three levels of performance. The four criteria provide a way to frame one's thinking as reflection takes place about what is important for students, teachers, and parents to know about a child's performance, achievement, and/or progress. These are the lenses through which teachers can view student responses to open-ended questions in mathematics. In alignment with the new ideas of assessment and based on the nature of intelligence and views of learning and teaching, we used the following "lenses" to frame the learning dimensions of open-ended questions in mathematics: conceptual, procedural, processing, and attitudinal. Details about these dimensions follow in the "Criteria" section.

As stated previously, there can be anywhere from two to ten levels of performance in a rubric. We chose to use rubrics with three levels of performance, where 3 indicates the highest quality. We found that it is often easier to focus first on what constitutes a Level 3 performance, then Level 2, and, finally, Level 1.

	Level 3	Level 2	Level 1
Conceptual			
Procedural			
Processing			
Attitudinal			

CRITERIA

The following is an explanation of each of the criteria sets, learning dimensions, or lenses through which open-ended questions in mathematics can be viewed.

- ◆ Conceptual Dimension

    This dimension includes mathematical concepts, ideas, principles, and theories. This is the "meat" of the mathematical content, looking at what the student knows. The major question that emanates from this domain is "Did the student understand the specific concept(s) involved in this question and, if so, where is the evidence that shows he/she understands it?"

- ◆ Procedural Dimension

    Procedures are ways to do things. They are the algorithms we learn (often by rote only) that allow us to accomplish tasks fairly efficiently and accurately. The key question in this dimension is "Did the student arrive at a correct solution and did he/she show evidence of using an appropriate algorithm and/or set of skills?"

- ◆ Processing Dimension

    This dimension incorporates processes that cut across all mathematical fields and topics. It hones in on assessing students' abilities to problem-solve, communicate, reason, make connections, and represent their thinking. The key question here can be multifaceted depending upon the open-ended question and what it entails and can, therefore, evolve into many questions:

    - • "Is there evidence of a specific problem-solving strategy?"

- "Did the student communicate his/her thinking?"
- "What is the evidence that supports the student used mathematical reasoning?"
- "Did the student show that he/she made the appropriate connections?"
- "Is there a representation of the student's thinking?"

♦ Attitudinal Dimension

This dimension often receives little attention in formal assessment practice. It is, however, a dimension that provides great insight into the student—insight that furnishes information to help maximize the child's learning. Some of the questions that fit into this dimension include:

- "What is the level of persistence, initiative, risk-taking, and creativity?"
- "How does the student work with others, listen to others, and interact with others?"
- "What is the student's level of curiosity about and enthusiasm for the mathematics in this question?"

EXAMPLE

One of the specific open-ended questions highlighted here follows an activity with base-10 blocks in which second-grade students were asked to build a structure, describe it, estimate its value, find the actual value, and show their work.

The building of the rubric proceeded through the following steps for each of the domains. The example provided details the "fleshing out" of the conceptual domain by asking and answering the following questions.

CONCEPTUAL DIMENSION

What was/were the major mathematical concept(s) involved in this task?

1. An understanding of the connection between digits and their values
2. Ability to bridge decade and century jumps when composing numbers

What defines level of performances?

♦ Concept 1: Connection between digits and their values

- Level 3—The student demonstrates an understanding of the connection between digits and their values (place value). This would be shown in the explanation provided by the student.

- Level 2—The student demonstrates some understanding—almost there but not quite.
- Level 1—The student demonstrates little or no understanding at all of the concept.

♦ Concept 2: Bridging decade and century jumps

- Level 3—The student demonstrates the ability to bridge decade and century jumps when composing numbers, for example, moving from 79 to 80, from 99 to 100 and 101, etc.
- Level 2—The student demonstrates some ability to make the jumps; may be able to make the decade jumps but unable to do so at the century mark.
- Level 1—The student demonstrates little or no ability to make the jumps and is unsure of how to proceed when faced with the task

The frame of the rubric thus begins to fill in:

	Level 3	Level 2	Level 1
Conceptual	• demonstrates understanding of connection between digits and their values (place value)	• demonstrates some understanding of connection between digits & their values (place value)	• demonstrates little or no understanding of connection between digits & their values (place value)
	• demonstrates ability to bridge decade & century jumps when composing numbers	• demonstrates some ability to bridge decade & century jumps when composing numbers	• demonstrates little or no ability to bridge decade & century jumps

This process should continue with the other learning dimensions—by identifying key questions within each dimension and then detailing what a response might look like at each of the three levels of performance. Questions for each of the remaining dimensions could be:

PROCEDURAL DIMENSION

What are the procedures that were used to complete the task?
Was the procedure completed accurately?

PROCESSING DIMENSION

Which of the processes were involved in responding to the following tasks: problem solving, communication, reasoning, connection making?

ATTITUDINAL DIMENSION

Which attitudes or dispositions could be or were displayed during this task completion: persistence, curiosity, enthusiasm, motivation, initiative, risk taking, ability to work collaboratively, and so on?

RUBRIC REVISION

It is important to note that it is difficult to construct a rubric that works "perfectly" the first time. In fact, revision is part of the standard operating procedure of rubric construction. All rubrics for open-ended questions need to be adjusted, tweaked, and massaged after responses to the question have been collected and initially analyzed using the rubric. This is something that should happen after the first use of the question and rubric and may continue to happen in subsequent use as both the question and rubric are refined and honed.

SUMMARY

The multiple purposes of assessment can cluster around creating opportunities to offer feedback for reflection and revision as the more generic purposes of assessment are served: to monitor students' progress, make instructional decisions, and evaluate students' achievement. The use of open-ended questions furnishes greater insight toward these goals if assessed in greater depth than being either just right or wrong. Open-ended questions can be evaluated either holistically or analytically depending upon the situation and the needs of the evaluator. A rubric—a set of criteria used to judge, evaluate, critique, and assess student work—can be utilized as a guide. Rubrics can be comprised of any number of criteria and have as few as two to as many as ten levels of performance. Rubrics can be constructed by teachers working as individuals, with other teachers, or with students, or by students working either individually or in small groups. Rubrics almost always need to be revised and refined after first and maybe even subsequent use.

A frame for rubric construction is offered that consists of four sets of criteria and three levels of performance. These criteria represent various learning dimensions in mathematics and provide a way to frame one's thinking about what is important for students, teachers, and parents to know about a child's performance, achievement, and/or progress. They are the lenses through which teachers can view student responses to open-ended questions in mathematics. They are in alignment with the new ideas of assessment and are based on the nature of intelligence and views of learning and teaching. The criteria focus on four identified learning dimensions: conceptual, procedural, processing, and attitudinal.

Rubrics can be constructed by examining a specific open-ended question and thinking of it in terms of each of the learning dimensions by asking

- ♦ What are the concepts addressed by this question?

- ♦ Which procedures can be used to complete the question?

- ♦ What are the processes that can or should be utilized?

- ♦ What are the attitudes that may be displayed while responding to the question?

6

WHERE
OPEN-ENDED
QUESTIONS
ARE FOUND

STANDARDIZED TESTS

One reason so much attention is being paid to the use of open-ended questions, sometimes referred to as constructed-response questions, is that most test and textbook publishers are including them as an option, if not the heart of their program. Those who do not are making plans to include them in future endeavors. The information given in this chapter is by no means exhaustive, and it is likely to become outdated quickly because change in mathematics education seems to be the norm of late. Be sure to call organizations and companies directly to inquire about their newest products.

STATE DEPARTMENTS OF EDUCATION

> As a major part of the educational reform effort of the 1990s, states have explored alternative forms of assessment, which require students to produce answers rather than simply select correct answers. Instead of replacing traditional assessments, however, most states are adding nontraditional assessments to their state assessment programs to measure those skills that cannot be measured with multiple-choice assessment....The most common forms of alternative assessment are paper-and-pencil extended and short answer written items on on-demand performance assessment. (Bond et al., 1997, p. 10)

This information is gleaned from the Council of Chief State School Officers' survey which is administered regularly to study the trends in state-sponsored assessment. With respect to mathematics, the report indicates that in 1997, 24 states were piloting or using nontraditional assessments and 16 more were developing such programs. For more information about the survey results, call the Council of Chief State School Officers at 202-408-5505, or e-mail edroeber@ aol.com.

A number of the state departments of educations (DOEs) have released samples of constructed-response questions that were used on state assessments, many of which are open-ended questions.

Kim's teacher asked her to use four colors to divide a square into parts, and to color the parts as follows:

   ½ is colored red → ¼ is colored blue

   ⅛ is colored green → any other part is left white

On squares below, draw three different designs to help Kim. Each square must contain all four colors.

On your sketches, write R for red, B for blue, G for green, and F for white.

Circle one of your designs and explain to Kim how you know that the parts you colored are the correct fractions.

The Floral Clock in Frankfort is being planted with flowers. About 4500 flowers have been planted so far.

- ♦ Estimate how many flowers will be on the Floral Clock when the gardeners have finished.

- ♦ Explain how you made your estimate.

OPEN RESPONSE 2

> **The fourth graders in Ms. Hartman's class bought some peanuts that come in small bags. Each student in the class reported how many peanuts were in her or his bag. Here are the numbers each student reported:**
>
> **Numbers of Peanuts in a Bag**
>
> **14 17 15 16 18 21 13 15 14 15**
>
> **15 16 17 17 19 17 16 19 15 19**
>
> **If someone asked you, "About how many peanuts are in a bag?" What would you say? Explain your thinking at each step and your answers.**

These samples are from California, Kentucky, and Oregon, respectively (National Council of Supervisors, *Mathematics,* 1996, pp. 2a–36, 2a–2, 2a–29). Information about other items and scoring rubrics is generally available from the DOE in most states. Some can be obtained through their Web sites. These can be reached by searching for the *state name* and "department of education."

NATIONAL AND INTERNATIONAL ASSESSMENTS

State initiatives are being influenced by two tests that have been given to samples of students across the country. The National Assessment of Educational Progress (NAEP), has been given roughly every four years since 1973. In the 1996 administration, more than 50 percent of the items were devoted to constructed-response questions. Some states include NAEP items on their own tests so that they can make national comparisons. The following three grade-four samples were taken from the Web site at: www.nces.ed.gov/ nationsreportcard.

> **Brett needs to cut a piece of string into four pieces without using a ruler or other instrument.**
>
> **Write directions to tell Brett how to do this.**

> **Think carefully about the following question. Write a complete answer. You may use drawings, words, and numbers to explain your answer. Be sure to show all of your work.**
>
> **The gum ball machine has 100 gum balls: 20 are yellow, 30 are blue, and 50 are red. The gum balls are well mixed inside the machine.**
>
> **Jenny gets 10 gum balls from the machine.**
>
> **What is your best prediction of the number that will be red.**
>
> **Explain why you chose this number.**

Generally students' performance on such constructed-response items has been poor. But there is hope. "As teachers reported increased emphasis on developing reasoning to solve unique problems, students' performance also increased on all forms of assessment items in NAEP" (Kenney & Silver, 1997, p. 23). This important finding suggests that increased use of open-ended questions at the classroom level should result in improved student performance on standardized tests.

Results of the *Third International Mathematics and Science Study (TIMSS)* also showed that American students have considerable difficulty with constructed-response questions. Only 16 percent of third graders and 24 percent of fourth graders responded correctly to the following relatively easy question (International Association for the Evaluation of Educational Achievement, 1997).

EXAMPLE

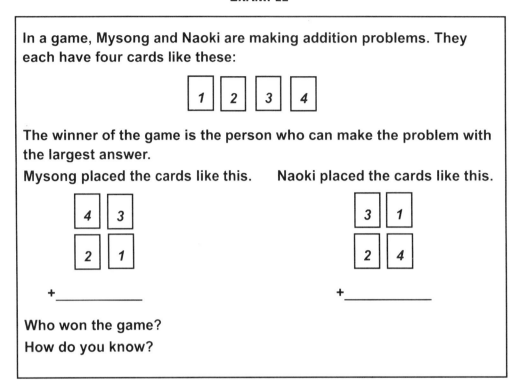

As with the state and national tests, TIMSS has recently been incorporating more open-ended type questions. One-fourth of the items, representing one-third of the testing time, was of the constructed-response type. All of the items that have been released are available in the publication *TIMSS Mathematics Items: Released Set for Population 1 (Third and Fourth Graders)* (International Association for the Evaluation of Educational Achievement, 1997). There is also a Web site that can be reached through the NAEP site mentioned earlier.

PUBLISHERS

Commercial test publishers are well known for their work with norm-referenced tests that have a multiple-choice, also called selected-response, format. The *www.heinemann.com* Web site states, "The primary purpose of many such tests is to rank-order students, their teachers, and their schools: that is, to guarantee that some will be labeled as successes and others failures, with the vast majority considered mediocre." Fortunately, tests publishers are now offering products that have other options such as sections of open-ended questions that can be reported out differently than the bell-shaped curve based on nationally normed data referred to in the Heinemann quote. The CTB/McGraw-Hill (1-800-538-9547) *Terra Nova* is one example. This test has both multiple-choice and open-ended questions. CTB/McGraw-Hill also publishes the norm-referenced CTBS and the CAT tests that are used by many school districts.

Riverside Publishing (1-800-323-9540) offers *Constructed Response Supplement and Performance Assessments* to accompany their norm-referenced *Iowa Test of Basic Skills*. Their *Multiple Assessment Series* is specifically for grades one through three and contains open-ended questions as does *Test for Success*, its test preparation product for grades K through eight.

Harcourt Educational Measurement (1-800-211-8378) publishes the norm-referenced *Metropolitan Achievement Test* (MAT) and the *Stanford Achievement Test* (SAT). Their ninth edition of the SAT includes an open-ended assessment booklet option. Their *Select* series offers locally customized options such as subtests with items added or deleted to meet a district's criteria. *Key Links* is one of their assessment support products. A workbook for each grade level provides many open-ended question experiences.

"The majority of states have worked with commercial test publishers on at least one statewide assessment. This seems to be done for two reasons: to insure technical quality when open-ended questions are included and to compare performance of their students with national normed-referenced tests results" (Bond et al., 1997, p. 7). Though much of this work relates to multiple-choice, norm-referenced testing, some states and commercial publishers are beginning to develop assessment programs that include questions that gather authentic student work. For example, Riverside Publishing has helped Michigan, Georgia, Ohio, and Washington develop criterion-referenced tests with open-ended components. Criterion-referenced tests are designed to assess students' performance with specified objectives or standards.

The *New Standards* is an interesting hybrid product. The *New Standards Project* produced a set of internationally benchmarked standards for student performance and an assessment system that measures performance against the standards. The examination portion for grades four, eight, and ten is published at Harcourt Brace Educational Measurement (1-800-211-8378). The standards themselves and the portfolio system are available from the National Center on Education and the Economy at the University of Pittsburgh (http://www.ncee.org).

FULL MATHEMATICS PROGRAMS

There are three elementary programs that have been developed with funding from the National Science Foundation. Though the programs are now distributed by commercial publishers, they were written by teams from research institutions. Generally, these programs contain more open-ended question opportunities than other full elementary mathematics programs. Some samples from fraction units at grade three follow.

SCOTT FORESMAN'S (1-800-552-2259)
INVESTIGATIONS IN NUMBER, DATA, AND SPACE PROGRAM
(TIERNEY & BERLE-CARMAN, 1998)

♦ "What are some things you notice about fractions?" (p. 9)

♦ "How can you combine two fractions to make a new fraction?" (p. 9)

♦ "How can you cut a fraction to make new fractions?" (p. 9)

♦ "Imagine that you have seven brownies to share equally among four people....See if you can find out exactly how many brownies each person will get." (p. 14)

♦ "Some people think 1 and ¼ is a larger share than 1 and ⅓.

Some people think 1 and ⅓ is a larger share than 1 and ¼.

Which do you think is bigger?

KENDALL HUNT'S (1-800- 542-6657)
MATH TRAILBLAZERS
TIMS, 1997

♦ "Draw a picture of people in your family. Then write several fractions for parts of your family. Explain each fraction. For example, if you live with your mother and two younger sisters, then you could write 1/4 for the fraction of your family that is grown up." (p. 568)

♦ "Below is one way to divide the rectangle in half. Find at least six other ways to divide the rectangle in half." (p. 571)

♦ Journal prompt: "Imagine you have divided something into two parts. What must be true about the area of the parts in order for them to be halves?" (p. 572)

EVERYDAY LEARNING'S (1-800-382-7670)
EVERYDAY MATHEMATICS
UNIVERSITY OF CHICAGO SCHOOL MATHEMATICS PROJECT, 1998

- "Write a short number story about something in the classroom. Use one or more fractions in your story." (p. 370)

- "Use your ruler or tape measure to measure three things at your desk or in the room. Use fractions or decimals, as needed, to record the measurements." (p. 373)

- "With a partner, sort the fraction cards in the *Everything Math Deck* by denominators. Make notes of anything you notice about fractions." (p. 379)

OTHER RESOURCES

PERIODICALS

- *Exemplars: A Teacher Solution* (1-800-450-4050) offers performance tasks for grades K through eight. Subscribers receive Fall, Winter, and Spring installments in hard copy and digital files. These include: questions/tasks, instructional tips, scoring rubrics, keys to national standards, and examples of student work.

- *Teaching Children Mathematics* is a National Council of Teachers of Mathematics journal published 10 times a year. It contains articles and activities in support of the latest initiatives in mathematics education, including assessment through open-ended questioning.

ACTIVITY BOOKS

CREATIVE PUBLICATIONS
(1-800-669-3986)

- *Puddle Questions*, by Joan Westley (1994), offers a book of activities for each grade, one through eight. These include instructional techniques, assessment criteria, and student work samples for at least eight open-ended questions. Each grade level opens with the same question, "How would you measure a puddle?" Some schools are collecting each year's response to this question in a student's progressive portfolio to show growth.

♦ *Connections: Linking Manipulatives to Mathematics*, by Linda Holden Charles and Micaelia Randolph Charles (1989), is an eight-book series that helps teachers pose many interesting problems in all domains of mathematics. Teaching and student recording tips are included with each activity.

♦ The *Constructing Ideas Series*, by Sandra Ward (1995), are activity books adapted from similar activities in the *Mathland Program* also from Creative Publications. There are three books for primary and three for intermediate, and each deals with a particular mathematics domain such as numeration, patterns, or operations. The lessons are clearly detailed and include teaching tips, ideas for "reflecting together," samples of student work, and home activities.

♦ *Writing Mathematics*, by Nancy Bosse (1995), is a series of teacher resource books available for grades one through six. Each book has 10 investigations that require students to collect data and report about things they have found. Also included are assessment criteria and many ideas about how to help students improve their craft.

♦ *Thinking Questions*, by Kathryn Walker, Cynthia Reak, and Kelly Stewart (1995), is a series of 15 books that focuses on students' use of manipulatives to solve problems. There are 20 main questions in each book with related discussion questions and journal prompts. *What You Might See, What to Look for In Student's Work*, and *What You Should Do If* are regular features of the teacher tips provided.

ADDISON WESLEY PUBLISHING
(1-800-848-9500)

♦ *Making Numbers Make Sense: A Sourcebook for Developing Numeracy*, by Ron Ritchhart (1994), offers 35 lessons for grades K through eight that employ open-ended investigations to support concept development and authentic assessment.

♦ *Kids Are Consumers, Too*, by Jan Fair and Mary Melvin (1986), is a collection of real-world mathematics activities. Some are lengthy projects; others are briefer opportunities to apply skills; and most integrate the applications of several mathematical domains and are appropriate for the intermediate grades.

CUISENAIRE DALE SEYMOUR PUBLICATIONS
(1-800-237-3142)

♦ Several different books by Marilyn Burns are good resources. Ms. Burns is a prolific author of materials and workshops that help teachers present mathematics with open-ended activities. Some titles are: *Math for Smarty Pants, A Collection of Math Lessons,* and the *Math by All Means Series: Place Value, Geometry, Probability, Multiplication, Division, Money,* and *Area & Perimeter.*

♦ *Family Math* by Jean Stenmark, Virginia Thompson, and Ruth Cossey, from the Lawrence Hall of Science at UC Berkeley, CA, is collection of games and activities designed to help parents and children work together to explore mathematical concepts. Many schools use these ideas to develop Family Math nights and home/school partnerships.

WEB SITES

Few sites have many open-ended questions. The ones given here are typical. These sites offer students interesting challenges that require thought but usually yield a single solution. These sites are changed often and it is hped that each will include more open-ended questions in the future.

♦ *http://forum.swarthmore.edu* is a site that offers many opportunities for teachers and students and links to other sites. Mathworld, Ask Dr. Math, and the Problem of the Week are student favorites at this site.

♦ *www.wonderful.com* is the Wonderful Ideas for Teaching, Learning, and Enjoying Mathematics site. The Kids' Corner brain teasers have some open-ended questions.

♦ *www.corona.bell.k12.ca.us/teach/swa/math.html* shows project ideas donc by grade five students during the last school year. It is an interesting model for students and teachers who are interested in sharing their work on the Web.

SUMMARY

The use of open-ended questions has grown tremendously in the last decade, particularly for state, national, and internationally sponsored assessment. Because we tend to learn what is tested, it is important to include open-ended questions during class instruction and during class testing. Students need experiences with open-ended questions at all grade levels. Traditional standardized tests do not permit sufficient opportunity for students to learn to construct robust written responses to problems.

Publishers of mathematics tests and mathematics materials are responding to this need by releasing sample test questions and activity books. These are good starters, but for assessment to fit instruction, teachers will also need to design some of their own questions. Review Chapter 3, "How to Create Open-Ended Questions," for suggested techniques.

7

EXAMPLES OF OPEN-ENDED QUESTIONS

This chapter provides a variety of open-ended questions in mathematics that have been used in elementary classrooms. The questions are grouped according to grade level spans. The primary grade level section targets kindergarten through grade two, and the intermediate level covers grades three through five. Within the grade level spans, the questions are grouped according to five basic mathematical strands:

- ◆ Number and Operations

- ◆ Patterns, functions, and Algebra

- ◆ Geometry and Spatial Sense

- ◆ Measurement

- ◆ Statistics and Probability

Each open-ended question sample has several components. There is a general information page for each where the topic, grade level, title, and question are identified. A classroom scenario is next. This provides a description of what took place in the classroom during the activity. Following this information is the rubric that was used to evaluate the papers, comments on how the sample papers were evaluated, and the papers themselves. You will notice that most of the rubrics follow the basic frame described in Chapter 5 using the four dimensions:

- ◆ Conceptual

- ◆ Procedural

- ◆ Processing

- ◆ Attitudinal

We believe that a consistent approach to rubric development helps students grasp what is important when writing a good response. Inviting students to participate in rubric construction is also an effective technique for this purpose. The last question, the grade five *Raisin Counts*, shows a sample of a class-constructed rubric that has been embedded in the four-section frame. Rather than specifying three levels of performance as the other examples do, the rubric shows a list of assessment criteria. Students find this analytic approach easy to use for peer and self-editing. They seem particularly motivated by the idea of earning more points through writing revisions. We have found that students in all grades can help to develop a rubric for most tasks, though fewer criteria than shown in the example is advisable for younger students.

PRIMARY GRADE SAMPLES

TOPIC: NUMBER AND OPERATIONS

GRADE: 2

TITLE: What's It Worth—Base 10

> QUESTION: What is the value of the structure you have built with base-10 blocks?

CLASSROOM SCENARIO:

This activity takes place after students have already explored with base-10 blocks and are fairly familiar with them. This was done with second graders near the middle of the year (early March) but can also be done with first-grade students toward the middle and end of the year.

Students were asked to build a structure that represents something seen or used in everyday life. A restriction was placed requiring that at least one of each type of block be used: unit, long, or flat. Each student was asked to provide a description of his/her structure. The second part of the activity focused on the value of the structure. Each student was asked first to estimate the value, next to find the actual value, and, finally, to show how he/she arrived at the final answer. This activity is one that would be a good portfolio component for showing progress over time when done at the beginning, middle, and end of the school year.

What's It Worth-Base 10

	Level 3	Level 2	Level 1
Conceptional	• demonstrates understanding of connection between digits and their values (place value)	• demonstrates some understanding of connection between digits and their values	• demonstrates little or no understanding of connection between digits and their values
	• demonstrates ability to bridge decade and century "jumps" when composing numbers	• demonstrates some ability to bridge decade and century jumps when composing numbers	• demonstrates little or no ability to bridge decade and century jumps when composing numbers
Procedural	• arrives at correct actual value	• arrives at correct or reasonable actual value	• arrives at incorrect actual value
	• fully and clearly demonstrates an addition procedure	• demonstrates an addition procedure that may be somewhat difficult to follow	• does not demonstrate an addition procedure or presents one difficult to follow
Processing	• provides reasonable estimate	• provides fairly reasonable estimate	• does not provide estimate or provides unreasonable one
	• utilizes an organized approach	• utilizes a moderately organized approach	• utilizes an approach with little or no organization
	• fully and clearly describes structure	• describes structure moderately	• describes structure inadequately
Attitudinal	• demonstrates persistence in task completion	• demonstrates some persistence with support	• demonstrates little or no persistence
	• demonstrates creativity in building and describing structure	• demonstrates some creativity in building and describing structure	• demonstrates little or no creativity in building and describing structure

What's It Worth? ~ Base 10

✓ Think of something that you see and/or use in everyday life.
✓ Make a structure that represents this item.
✓ Use some units, some longs, some flats.
✓ Describe using words and/or pictures.

✓ Estimate the value of your structure.
✓ Put a line around it.
✓ Find the actual value.
✓ Tell/show how you arrived at your answer.

I made a man Walking down the sidewalk. And the sun is very bright on that day. It U.a saturday morningh and the man Is exersizing

5o9

3 100's

25 10s

7 1s

3 100's = 300
25 Tens = 250 (when ever you have a number moltaplied by 10 you add a zero)
7 1s = 7

300 + 200 = 500
500 + 50 = 550
550 + 7 = 557

Level 3

In the Level 3 response, the student clearly demonstrates an understanding of the connection between digits and their values. It is also clear that the student can make both decade and century jumps. The addition procedure is easy to follow. The estimate is reasonable and the actual value is correct. An added bonus is the student's explanation of how to multiply by 10.

What's It Worth? ~ Base 10

Think of something that you see and/or use in everyday life.
Make a structure that represents this item.
Use some units, some longs, some flats.
Describe using words and/or pictures.

Estimate the value of your structure.
Put a line around it.
Find the actual value.
Tell/show how you arrived at your answer.

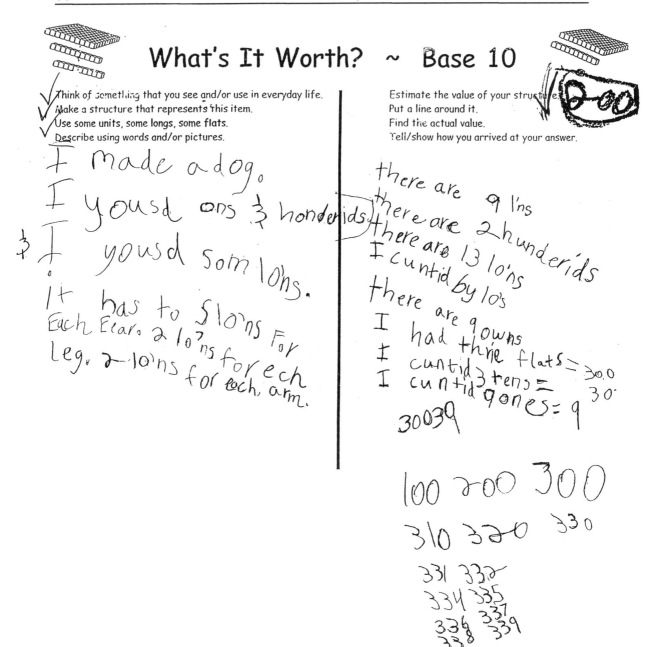

I made a dog.
I yousd ons & honderids
I yousd som lohns.
It has to 5 lohns For
Each Ear. 2 lohns for ech
leg. 2 lohns for ech, arm.

there are 9 lins
there are 2 hunderids
there are 13 lohns
I cuntid by 10's
there are 9 owns
I had thrre flats = 300
I cuntid 3 tens = 30
I cuntid 9 ones = 9
30039

100 200 300
310 320 330
331 332
334 335
        337
336   339
338

Level 2

The Level 2 response shows the ability to bridge the century jump from 200 to 300. The estimate could be more reasonable. The student demonstrates a procedure that is a difficult to follow, but the added explanation shows some organization. Interesting to note is the difference between the two representations.

What's It Worth? ~ Base 10

Think of something that you see and/or use in everyday life.
Make a structure that represents this item.
Use some units, some **longs**, some **flats**.
Describe using words and/or pictures.

Estimate the value of your structure.
Put a line around it.
Find the actual value.
Tell/show how you arrived at your answer.

I mad a garden with some seedlings and some flowers and a gate around it.

[139]

26 one

4 tens

1 100

I contid

31

Level 1

The Level 1 response indicates little or no understanding of the connections between digits and their values. This is underscored by the discrepancy between the estimate and actual value. There is no addition procedure used, and the actual value is incorrect.

TOPIC: PATTERNS, FUNCTIONS, & ALGEBRA

GRADE: 1

TITLE: Heads, Tails, and Other Such Things

> QUESTION: How many heads, tails, and _____ are there?

CLASSROOM SCENARIO:

A simple and engaging story was read to a classroom of first-grade students. *I Love You, Blue Kangaroo* (Clark, 1999) revolves around the main character, Lily, who falls asleep each night with her special kangaroo wrapped in her arms. Lily begins to accumulate more stuffed animals that, one by one, also begin to share her bed, much to Blue Kangaroo's consternation. The story is lovely. It is one that is rich with emotion and feelings common to first graders, and it provides wonderful mathematical opportunities. One such opportunity presents itself in the picture with Lily and the eight stuffed animals in a very crowded bed. A question that could emerge is how many heads and long tails are there. The "other such things" leaves open a variety of options. In this case, eyes were chosen so that students could determine how many heads, tails, and eyes were there among all of the animals and Lily. Students were allowed to work either individually or with a partner, and were asked to show their thinking with words, pictures, numbers, and/or other things. A variety of manipulatives were also made available to the students (cubes, tiles, etc.).

Heads, Tails, and Other Such Things

	Level 3	Level 2	Level 1
Conceptual	• demonstrates understanding of correspondence of heads & eyes to bodies	• demonstrates understanding of correspondence of heads & eyes to bodies	• demonstrates some or little understanding of correspondence of heads & eyes to bodies
	• applies knowledge of doubles	• may/may not apply knowledge of doubles	• does not demonstrate knowledge of doubles
Procedural	• arrives at correct solution	• arrives at correct or reasonable solution	• arrives at incorrect solution
	• fully demonstrates an addition procedure	• partially demonstrates an addition procedure	• does not demonstrate an addition procedure
Processing	• provides clear representation of strategy	• provides partial representation of strategy	• provides little or no representation of strategy
	• utilizes an organized approach	• utilizes a moderately organized approach	• utilizes an approach with little or no organization
Attitudinal	• demonstrates persistence in task completion	• demonstrates some persistence with support	• demonstrates little or no persistence
	• works well with a partner	• works satisfactorily with a partner	• does not work well with a partner

Heads and Tails and Other Such Things

We figured out how many **heads, tails,** and _eyes_____ there were in the problem.   This is how we solved it:

I lookd at the anim
als and Lily. I got 9.

Every Head has 2 eyes
so, if you count by 2s you
will end up on 18.

The Blue Kangaroo
has a long tal. The
Wiggly green Crocadile has
a long tal and the 2
pupies have tals,

It all two gather
is 31.   18 + 9 + 1 + 1 + 2
31. 18 + 9 = 27, 27 + 4 = 31.

Level 3

   The organized approach demonstrated in the Level 3 response begins with the correct correspondence among the characters, their heads, and their eyes. The student then clearly applies and communicates knowledge of doubles. The addition procedure is presented in an overall fashion (18 + 9 + 1 + 1 + 2 = 31) and then backed up by a look at its components (18 + 9 = 27 and 27 + 4 = 31).

Level 2

In the Level 2 response, the student shows an organized approach to solving the problem by determining the number of tails according to character, the number of eyes by listing an initial for each character, and then doubling or counting by twos. The student arrives at the correct solution, and even though we know she "kotab" (counted), the strategy/procedure used is not clearly represented.

Level 1

Although the Level 1 response does reflect an understanding of the correspondence of heads and eyes to bodies, there is no representation of the strategy used to solve the problem which is solved incorrectly.

TOPIC: GEOMETRY AND SPATIAL SENSE

GRADE: 2

TITLE: Sorting Solids

QUESTION: Explain in sentences how different prisms are alike and how they are different. Explain in sentences how spheres and cylinders are alike and how they are different.

CLASSROOM SCENARIO:

This written work followed a lesson where groups of four students were asked to sort a set of 16 wooden solids into different sets so that like things were together. The other students were invited to view each group's sort and to guess the attributes of each of the subsets. The creators of the subsets then gave their own interpretation.

Rubric	Level 3	Level 2	Level 1
Conceptual	• description includes geometric characteristics such as roundness or flatness	• description lacks geometric attributes	• no attributes are included
Processing	• explanation is logical and understandable	• explanation needs better organization	• explanation does not to relate to the question
Attitudinal	• response addresses both parts of the questions	• response only addresses part of the question	• student was not on task and did little work

1. **Explain in sentences how prisms are alike and how they are different.**

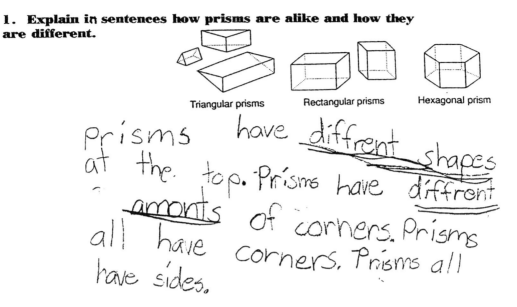

Triangular prisms   Rectangular prisms   Hexagonal prism

Prisms have diffrent shapes at the top. Prisms have diffrent amonts of corners. Prisms all have corners. Prisms all have sides.

2. **Explain in sentences how spheres and cylinders are alike and how they are different.**

Spheres   Cylinders

Spheres are like a ball. Cylinders are like a trunk of a tree. Cylinders don't have corners. Cylinders don't have sides. Neither do Spheres

---

Level 3

The Level 3 paper identifies geometric characteristics such as roundness and corners. The explanations are understandable to others. All parts of the two questions are answered. Question one is particularly well done.

1. Explain in sentence ✗ how prisms are alike and how they are different.

Triangular prisms   Rectangular prisms   Hexagonal prism

*theay each have
rektagals / sqares.
Some have triag ils some are
Jeast a litil rownd,*

2. Explain in sentence ✗ how spheres and cylinders are alike and how they are different.

Spheres   Cylinders

*theay all dont
have corners.*

Level 2

The Level 2 paper is not complete, having only one statement for question two. The expression "jeast a litil rownd" lacks clarity in geometric terms.

1. Explain in sentences how prisms are alike and how they are different.

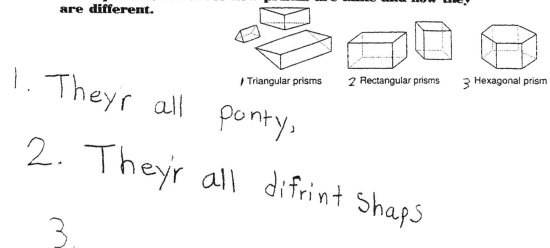

1. Triangular prisms   2 Rectangular prisms   3 Hexagonal prism

1. They'r all ponty,

2. They'r all difrint shaps

3.

2. Explain in sentences how spheres and cylinders are alike and how they are different.

Spheres   Cylinders

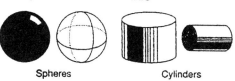

Level 1

The Level 1 paper is not complete because only part of question one has been answered. Even that response lacks geometric clarity.

TOPIC: MEASUREMENT

GRADE: 2

TITLE: Popsicle Perimeter

> QUESTION: How many different closed shapes can you make with a certain number of popsicle sticks? What do you notice?

CLASSROOM SCENARIO:

This was an interesting activity for second graders at the end of the year but can also be used with older students. The second graders had a mini-lesson beforehand that modeled a task similar to the one they would be asked to do. The class was presented with the task of making a closed shape using eight popsicle sticks. Once one was made, a volunteer was asked to record it on the overhead on graph paper and label the perimeter. The next challenge was to make a different shape with the same perimeter, then record and label it. This continued until the students had made all of the different shapes they thought they could make.

Following the class lesson, the students were asked to work with a partner to answer the question about how many different closed shapes they could make using 12 popsicle sticks (each student had his/her own recording sheet). They had to record their work on graph paper and label each shape with the perimeter. Students were asked to reflect upon their work and then write about relationships, connections, or anything interesting they may have noticed about their work.

Popsicle Perimeter

	Level 3	*Level 2*	*Level 1*
Conceptual	• demonstrates complete understanding of terms (closed shape & perimeter) and task	• demonstrates at least partial understanding of terms (closed shape & perimeter) and task	• demonstrates little or no understanding of terms (closed shape & perimeter) and task
	• understands how shapes with equal perimeters can have different areas	• may understand how shapes with equal perimeters can have different areas	• has little or no understanding that shapes can have equal perimeters but different areas
Procedural	• constructs, records, labels shapes with perimeters of 12	• constructs, records, labels shapes with perimeters of 12	• may/may not construct, record, label shapes with perimeters of 12
	• records >8 correct and different shapes	• records 5–8 correct and different shapes	• records <5 correct and different shapes
Processing	• demonstrates understanding of how to solve problem	• demonstrates satisfactory understanding of how to solve problem	• demonstrates little or no understanding of how to solve problem
	• clearly communicates observations including connection between perimeter & area	• partially communicates observations possibly with mention of connection between perimeter & area	• inadequately communicates observations with no mention of connection between perimeter & area
Attitudinal	• demonstrates persistence in task completion	• demonstrates some persistence with support	• demonstrates little or no persistence
	• works well with a partner	• works satisfactorily with a partner	• does not work well with a partner
	• displays willingness to take risks	• displays some willingness to take risks	• displays little or no willingness to take risks

POPSICLE PERIMETER

I noticed some pieces could fit together like a puzzle and I also noticed that even though the perimeter was 12 the area is differnt ... and a lot of the areas were five. Some of them look like steps. Some of the shapes looked like letters.

I thought it was going to be a piece of cake but it turned out to be a little hard because a cupple of times I mad the same shap but in a different direction.

One of the pieces that looked like a letter looked like an M and it also looked like steps.

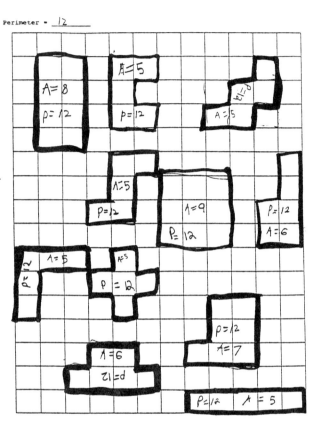

Perimeter = 12

| Level 3 |

In the Level 3 paper, the student has clearly shown understanding of how shapes can have the same perimeter but different areas. Evidence of the student's interest is highlighted by the remark about finding shapes with a perimeter of 12 that have an area of 5. She demonstrates persistence in completing the task which she thought at first would be a "piece of cake."

POPSICLE   PERIMETER

√~   Use popsicle sticks.

√~   Make a CLOSED SHAPE using all of the sticks.

√~   Record how many sticks you used on the graph paper.

√~   Record your shape on the graph paper.

√~   Make as many different shapes as you can using the same number of popsicle
        sticks. Record each shape.

~   Write about what you noticed.

We noticed that even if the perimeter is the same it doesn't mean
the area has to be the same. I mean look at these

this activity was hard in a way and was fun. P=12 P=12
It was hard finding the closed   A=6 A=5
shapes because we kept on doing the
same shapes as we did before and it
was fun.

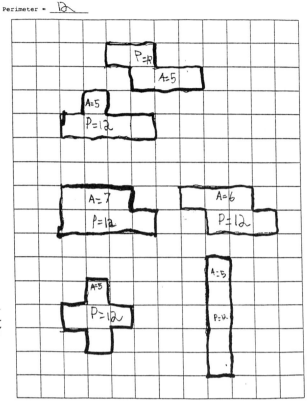

Perimeter = 12

Level 2

The Level 2 paper communicates an understanding of how shapes with the
same perimeter can have different areas, and it provides a drawing to back up
the thinking. It identifies six shapes and demonstrates a satisfactory understanding of how to solve the problem.

POPSICLE PERIMETER

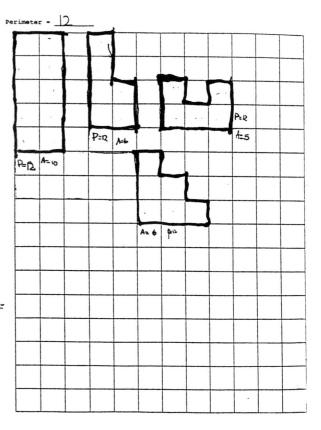

Perimeter = 12

- ✓ Use popsicle sticks.

- ✓ Make a CLOSED SHAPE using all of the sticks.

- ✓ Record how many sticks you used on the graph paper.

- ✓ Record your shape on the graph paper.

- ✓ Make as many different shapes as you can using the same number of popsicle sticks. Record each shape.

- ~ Write about what you noticed.

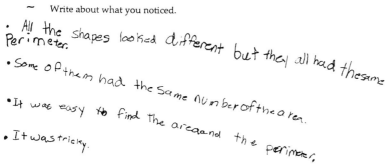

- All the shapes looked different but they all had the same Perimeter.

- Some of them had the same number of the area.

- It was easy to find the area and the perimeter.

- It was tricky.

Level 1

The response in the Level 1 paper does show an understanding that shapes can have the same perimeter but different areas, yet it shows lack of success in finding more than four shapes with a perimeter of 12.

TOPIC: STATISTICS AND PROBABILITY

GRADE: Kindergarten

TITLE: I Spy...

> QUESTION: What do you spy in this picture and how many things in all?

CLASSROOM SCENARIO:

An activity such as this one is very appropriate for kindergarten. It can be made to fit with any unit that is ongoing and can serve many purposes. All that is needed is multiple copies of an interesting picture that contains a variety of objects that can be sorted and classified. This picture can be from a favorite book, a placemat, a puzzle, etc. This is also an activity that can and should be done throughout the school year. It has been used successfully in October, November, January, April, and June. The samples that follow were from a group working in early April.

Students were gathered in small groups of four to six at a time. Each had a copy of a picture from *On Market Street* (Bunting, 1996). In the small group setting, students were asked what they spied in the picture. Some of the answers included: people, candy, and animals. They were then asked to think of some ways they could show on paper what they saw in the picture. They decided that they could use pictures, letters, words, and, of course, numbers! They started the task knowing that there would be another question at the end. This question was revealed after they were finished recording. The students were asked if they could think about how many items in all they showed on their papers. Some were able to show their thinking independently; others dictated their thoughts.

I Spy...

	Level 3	*Level 2*	*Level 1*
Conceptual	• demonstrates understanding of classification as a means of organizing data	• demonstrates some understanding of classification as a means of organizing data	• demonstrates little or no understanding of classification as a means of organizing data
	• demonstrates understanding of the idea that finding the total number of items in a large group requires breaking down smaller groups	• demonstrates some understanding of the idea that finding the total number of items in a large group requires breaking down smaller groups	• demonstrates little or no understanding of the idea that finding the total number of items in a large group requires breaking down smaller groups
Procedural	• counts accurately displaying 1-to-1 correspondence	• counts fairly accurately displaying 1-to-1 correspondence	• counts inaccurately without 1-to-1 correspondence
	• utilizes an addition strategy beyond counting objects	• utilizes addition strategy of counting each set of objects	• does not utilize an addition strategy
Processing	• provides clear and full representation of data	• provides partially clear and satisfactory representation of data	• provides unclear and inadequate representation of data
	• articulates clearly addition strategy	• articulates satisfactorily addition strategy	• does not articulate addition strategy
Attitudinal	• demonstrates persistence in task completion	• demonstrates some persistence with support	• demonstrates little or no persistence
	• displays excitement about and interest in the task	• displays some excitement about and interest in the task	• displays little or no excitement or interest in the task

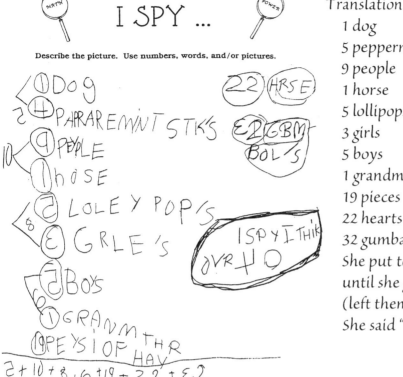

Translation of Student Work
1 dog
5 peppermint sticks
9 people
1 horse
5 lollipops
3 girls
5 boys
1 grandmother
19 pieces of hay
22 hearts
32 gumballs
She put together 2 numbers at a time
until she got to the last 3 big numbers
(left them alone)
She said "I spy...I think...over 40"

Level 3

The Level 3 response indicates an understanding of how items can be classified in more than one way; nine people are reclassified as boys, girls, and a grandmother. This is a full and clearly depicted representation of the data because of student persistence, excitement, and interest. An obvious bonus is the information gained about her thinking when she responds to "How many in all?" She puts together two numbers at a time until she comes to the last three large numbers and then writes a number string. She realizes that she cannot give the total yet but knows that it has to be "over 40."

Translation of Student Work

11 gumballs
1 dog
1 horse
5 lollipops
1 mouse

How many in all?
(counted 11 gumballs one by one —
then the 5 lollipops — then the
singles) = 18

Level 2

_____

The Level 2 response depicts a satisfactory representation of the data with accurate counting and one-to-one correspondence. In response to "How many in all?", the student utilizes the addition strategy of counting from the largest number, starting with counting the 11 gumballs one by one, then doing the same with the 5 lollipops, and then the single items.

I SPY ...

Describe the picture. Use numbers, words, and/or pictures.

Translation of Student Work

10 gumballs
5 swirly lollipops
1 dog
5 little hearts
1 cup
1 horse

I don't know how to equal!

Level 1

In the Level 1 paper, the student counts inaccurately in two of the groups depicted. She demonstrates little persistence in providing an adequate representation of the data available. There is no articulation of an addition strategy. When asked "How many in all?", the response, "I don't know how to equal," was given.

INTERMEDIATE LEVEL

TOPIC: NUMBER AND OPERATIONS

GRADE: 3

TITLE: Adding Coins

> QUESTION: Mariella just doesn't seem to get the adding of two-digit numbers. Her teacher suggested that she use play money to try to figure out:
>
> $.46 + $.75
>
> Use pictures of coins and describe in words what she needs to do to get a correct solution.

CLASSROOM SCENARIO:

This was the students' second question of this type. From the prior work, it was apparent that the students needed to have manipulatives available to answer such a question. Because coins were limited, students worked in groups of four to act out how the addition could be done. The first student placed coins for $.46 on a money place value chart; the next, the $.75. The third student regrouped the pennies; the fourth, the dimes. Then all students worked independently to describe how the solution was found.

Rubric	Level 3	Level 2	Level 1
Conceptual	• explanation shows understanding of the regrouping process	• explanation is incomplete	• explanation has no information about regrouping
Procedural	• solution is correct	• solution is partially correct	• solution is not in the ball park
Processing	• explanation of how the work was done is logical	• explanation has some appropriate elements	• explanation does not relate to the problem
Attitudinal	• response shows persistence and is complete with pictures and words	• response does not include pictures or does not include words	• student was not on task and did little work

ADDING COINS

> QUESTION: Mariella just doesn't seem to get the adding of two-digit numbers. Her teacher suggested that she use play money to try to figure out:
>
> $.46 + $.75 ·
>
> Use pictures of coins and describe in words what she needs to do to get a correct solution.

take 4 dimes and take
6 pennies. Take 5 pennies and 7 dimes
The you add all the dimes together
just count by tens. then tack
ten dimes and trade it in for
1 Dollar. then count the pennies. trade
10 pennies for 1 dime. Then you would
have 1 Dollar and 2 dimes and one pennie.

Anser $1.21 ✓

Level 3

The Level 3 paper meets all the criteria, including a logical sequence, pictures, and a correct solution. The explanation of trading/regrouping is particularly well done for the grade level.

ADDING COINS

QUESTION: Mariella just doesn't seem to get the adding of two-digit numbers. Her teacher suggested that she use play money to try to figure out:

$.46 + $.75 ·

Use pictures of coins and describe in words what she needs to do to get a correct solution.

1.21

$$
\begin{array}{r}
46 \\
+ \;.75 \\
\hline
\end{array}
$$

You should add 6+5 **are the ones.** then add 4+7 in the tens. You should figure out that its in the dimes, but also it is low in the dollars. You would have 4 dimes and 6 pennies for 46 and 7 dimes and 5 pennies, You would come up with 1 dollar 2 dimes 1 pennie. It would equal $1.21.

Level 2

The Level 2 paper has some appropriate elements such as a correct answer, but the ordering of explanation statements could be done more logically. There is no mention of regrouping in the written explanation except for the one written above the dime's column.

ADDING COINS

QUESTION: Mariella just doesn't seem to get the adding of two-digit numbers. Her teacher suggested that she use play money to try to figure out:

$.46 + $.75 ·

Use pictures of coins and describe in words what she needs to do to get a correct solution.

Take 6 pennies and after take 4 dimes.
Then take 7 dimes then add then

Level 1

The Level 1 paper shows little effort and is far from complete. The student was not on task as evidenced by the fact the second sentence was not even finished.

TOPIC: PATTERNS & FUNCTIONS

GRADE: 4

TITLE: Making .25 With a Calculator

> QUESTION: Besides ¼, what other fractions will yield
> .25 on a calculator? Can you describe a pattern?

CLASSROOM SCENARIO:

This was the kickoff to a fraction unit. The teacher offered no assistance other than a review of how to enter a fraction on a four-function calculator. Students worked in groups of three to explore this question. They were assigned a specific role in the group. One student was directed by the others to try different values with a calculator. Another student recorded on chart paper the successful values and an explanation of the patterns they observed. The third student presented the group's findings to the class.

RUBRIC:

In this case no real rubric was created because the teacher was using this activity to inspire interest in equivalent fractions rather than to do individual assessments. All groups did perform successfully at Level 3, which might look like the following:

Rubric	*Level 3*
Conceptual	• presentation shows understanding that there is an infinite set of fractions equivalent to .25
Procedural	• solutions are all numerically correct
Processing	• explanation of how a series of fractions is produced is logical
Attitudinal	• student performs the group role well

The following samples of the student work were copied from presentation charts. All the groups were able to produce Level 3 responses. This shows how students working together can offer a good opportunity to model strong responses. This approach does not give a clear picture of the contribution of each individual student, however. A more informative approach would be to allow students to do the work together but to write up reports individually.

Making .25

Beside 1/4, what other fractions will yield .25 on a calculator?
Describe what you notice about how to find more factions.

$$\frac{1}{4} \quad \frac{2}{8} \quad \frac{4}{16} \quad \frac{8}{32} \quad \frac{16}{64} \cdots$$

We doubled both the numerator and denominator. They all equaled .25.

$$\frac{1}{4} \quad \frac{10}{40} \quad \frac{100}{400} \quad \frac{1000}{4000} \cdots \quad \frac{1,000,000}{4,000,000} \cdots$$

We added a zero to the top and bottom numbers. You could go on forever doing the same thing.

$$\frac{1}{4} \quad \frac{2}{8} \quad \frac{3}{12} \quad \frac{4}{16} \quad \frac{5}{20}$$ on and on.

The top numbers go by 1's, that is, 1, 2, 3, 4, 5. You add a 4 to each bottom number, that is, 4, 8, 12, 16, 20.

TOPIC: GEOMETRY AND SPATIAL SENSE

GRADE: 3

TITLE: Twins

> QUESTION:
>
> ♦ **What directions did you give to your partner to make the twin of your structure?**
>
> ♦ **Which parts of this activity were easy and which were hard?**

CLASSROOM SCENARIO:

The complete activity "twins" takes some time to build up to and requires some prerequisite work. Students must be familiar with pattern blocks prior to doing this activity. In addition, students should do the basic activity orally more than once before being asked to do any writing. The complete activity was done with third graders toward the middle of the year but can also be done with students in older grades.

After developing familiarity with both pattern blocks and the basic activity, the third-grade students were asked to do the complete twins activity. Students worked with partners and a set number of pattern blocks (a range of 10 to 15 works well for initial work). Each worked behind a shield, made a flat structure of some kind, and then recorded his/her work. Each student then wrote the directions to be given to his/her partner, with the goal of enabling the partner to build a twin of the structure being described. Partners then took turns reading their directions and comparing the resulting structures. A reflective piece was also required.

Twins

	Level 3	*Level 2*	*Level 1*
Conceptual	• exhibits complete understanding and use of geometric and spatial terms	• exhibits partial understanding of geometric and spatial terms	• exhibits little or no understanding of geometric and spatial terms
Procedural	• provides sequenced, organized directions that could lead to exact "twin"	• provides directions with some ambiguities that could lead to reasonably close "twin"	• provides directions that lack sequence and organization not likely leading to a "twin"
Processing	• writes directions that include all relevant information and appropriate level of detail	• writes directions that include some relevant information and satisfactory level of detail	• writes directions that include little or no relevant information with lack of detail
	• constructs drawings that match actual structures	• constructs drawings that are reasonable matches to actual structures	• constructs drawings that are inadequate matches to actual structures
	• clearly communicates areas of strength/difficulty in task completion	• satisfactorily communicates areas of strength/difficulty in task completion	• inadequately communicates areas of strength/difficulty in task completion
Attitudinal	• demonstrates persistence in task completion	• demonstrates some persistence with support	• demonstrates little or no persistence
	• works well with a partner	• works satisfactorily with a partner	• does not work well with a partner
	• displays willingness to reflect upon and share areas of difficulty	• displays some willingness to reflect upon and share areas of difficulty	• displays unwillingness to reflect upon and share areas of difficulty

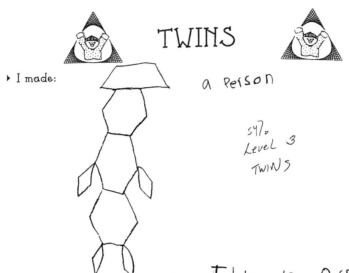

TWINS

▸ I made: a Person

54%
Level 3
TWINS

◆ Directions I gave to my partner: This is a Person.
you need 3 yellow hexagons, 4 white
Romby, and 1 Red trapezoid. First
Put the 3 hexagons and conect them
going down in a strate line* Then Put
the long side of the trapezoid
as a hat on the top of the top hexago.
Then take two of the Romby -make
them as the arms in the middle of the
middle hexagon and make them
in a down slant. Then make the
other two Romby as the feet. Put
them at a down slant on the tip
of the bottom hexagon.

*flat side to flat side.

◆ My partner made:

It was the same.

◆ We thought:

We thought it was hard to write the directions. But Easy to follow there directions. I found I had to add more detales about how to put the hexagons together. my partner had to gess. It was a little Easy to to do. because this was a Egsy dising. But it was hard to tell what angle to put the block on.

Level 3

The Level 3 response includes geometric terms that indicate an understanding of their use. The directions are clear, sequenced, and organized. The drawings match the actual structures that were built. The students were able to clearly communicate their areas of strength and difficulty. It is interesting to note their observation that their success may be due to the fact that theirs was an "easy design."

TWINS

‣ I made:

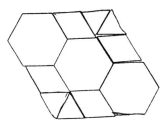

◆ Directions I gave to my partner:

3 ⬡

4 blue ◇

4 △

1. First take the 3 ⬡'s. Put the ⬡s

It looks like a space ship. So it looks like a yellow catipillar.

2. Then take the ◇ and put them in the spaces that are left.

3. Then take the △ and make 2 rombi. Then put them in the other 2 spaces.

◆ My partner made:

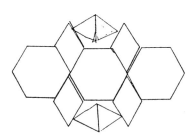

◆ We thought: I needed to explain the pasitions of the hexagons better. My partner said I should explain it better next time.

Level 2

The Level 2 response includes some relevant information with a satisfactory level of detail but also some ambiguities, for example, "put the hexagon so it looks like a yellow caterpillar." The drawings are reasonable matches. There is satisfactory communication of areas of strength.

TWINS

◆ I made ... a star

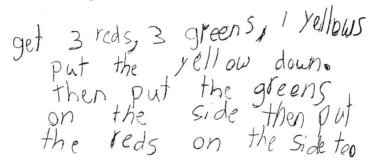

◆ Directions I gave to my partner:

get 3 reds, 3 greens, 1 yellows
put the yellow down.
then put the greens
on the side then put
the reds on the side teo

◆ My partner made:

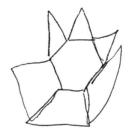

◆ We thought:

we thought it was confusing
because the directions were
hard to do

Level 1

The Level 1 response lacks details, including spatial terms. The drawings do not adequately match the actual structures. The student does communicate that the task is "confusing" and the directions "hard to do," yet does not indicate what could improve the work.

TOPIC: **MEASUREMENT**

GRADE: **3**

TITLE: **A Measurement Handbook**

> QUESTION: Someday life forms from other galaxies may visit us! It is likely they will measure things differently than we do.
>
> Use pictures and words to develop a handbook that shows how people on earth measure the length of things that are both short and long.
>
> Begin by listing important mathematical words that you are likely to talk about in the handbook, such as ruler, centimeters, length....
>
> You can put these in an idea web or in sentences on a new sheet of paper for your first draft.

CLASSROOM SCENARIO:

The students were in groups of three for this work. Because this was such a large task, the students and the teacher decided to break the handbook into three chapters with each group writing their own handbook. The chapter titles were:

- ♦ Why We Measure
- ♦ How We Measure
- ♦ What We Measure IN and WITH

At least one day was spent collecting important terms on the board. The terms were sorted by the chapter headings, and each student in a group was assigned one of the chapters to write.

SAMPLES:

The following samples are from the "How We Measure" chapter. These were submitted after the groups did some peer editing but before students saw a rubric.

Rubric	Level 3	Level 2	Level 1
Conceptual	• explanation shows a good sense of the need to place and read the measuring tool accurately	• explanation shows some confusion or an incomplete explanation	• explanation misses the point
Procedural	• information given is accurate	• information is partially correct	• response is not correct
Processing	• explanation makes sense and is easy to follow	• explanation needs better organization	• explanation does not relate to the question
Attitudinal	• response shows more than one example or idea	• response would be better if it included more information	• student was not on task

MEASUREMENT HANDBOOK

Some day we may have visitors from other galaxies!!
It is likely they measure things differently than we do.

♦ Use pictures and words to develop a handbook that shows how people on earth measure the length of things that are both short and long.

♦ Begin by listing below important words that you are likely to talk about in the handbook, such as ruler, centimeters, length…

♦ You can put these in an idea web or in sentences on a new sheet of paper for your first draft.

When you measure something you must measure from one point to another. When you measure a shoe you measure from heel to toe then you must Look at the number you got from the ruler. Or you can see how much and inch or a foot is and then estimate on your shoe. 1# You must start measuring from the one on the ruler 2# make sure your using the right kind of measurement.

Level 3

The Level 3 responses shows concern for accurately placing the ruler and measuring the entire length. The directions are clear and accurate, though they should be amended to say "start measuring from the zero" rather than from the one. The response only shows one example, but it is well done.

How to Measure

The way We measure is We estimate, or We put the tool we are measuring with agenst our object. We Also use numbers to measure the width and the length. The width is from side to side, and the length is from top to bottem.

Level 2

The Level 2 response exhibits a good start with the explanation about placing the tool "agent" the object. Unfortunately the work is not complete.

How TO Measure

If you are learning about measuring you might want to read this.

Inches: Inches

Centimeters: Centimeters is those things inside of the Inch ||||||.
cem.

Level 1

The Level 1 paper misses the topic to be discussed and gives incorrect information about centimeters being inside of inches.

TOPIC: STATISTICS AND PROBABILITY

GRADE: 5

TITLE: Raisin Counts

QUESTIONS: Look at the raisin data on the board and answer the following questions. Be sure to explain how you arrived at your responses.

- ◆ What is/are the mode(s)?

- ◆ What is the median?

- ◆ If the raisins were collected altogether and redistributed back into the boxes so that each box had the same amount, how many would be in each box?

- ◆ What trends do you see in the shape of the data?

CLASSROOM SCENARIO:

Each student counted the contents of a .5 oz. box of raisins and put the results on a sticky-note line graph. The students individually wrote responses to the questions and put them into a report.

The teacher and students created the rubric together. A student's work was first evaluated using the rubric by another student and then by the teacher. The attached reports are the second drafts following the student evaluations. Note: It was possible to exceed 12 points if a report contained significant extra information, such as the range.

Rubric	*One point for each thing the report includes:*
Conceptual (4 pts.)	• how you got the mode • how you got the median • how you got the mean (equal amount in every box) • why you think your trend is reasonable
Procedural (5 pts.)	• correct mode • correct median • correct mean • a trend • the graph
Processing (2 pts.)	• clearly written explanations • a description of the investigation (added to the set of questions after class agreement it was important)
Attitudinal (1 pt.)	• finished work

Report

We made the line plot by the teacher giving us a box of raisins. We counted the number of raisins in our boxes, and each person put a post-it on a number of raisins on the line plot. After everyone put a post-it on, we had a line plot of the number of raisins in our boxes.

In this lesson, I learned something new to graph in a line plot, what statistics are, raisins can have a big range, what trend means, and that our estimates for the number of raisins was off. This is because the raisins were smushed together.

The mode for our data was 31, the medium was 31, and the mean is somewhere between 30 and 31, our range was 26-35. Also, the trend of our data was probaby was that I noticed more people had post-its on the end of the data.

I got the mode by finding the most frequent value, I got the median by finding the lowest and highest numbers and counting til they met. I found the mean by adding the number of raisins and dividing by the number of raisins and dividing by the number of boxes. I found the trend by just looking at the line plot and that showed me that the ending numbers were more frequent than the beginning numbers.

Level 3

The Level 3 report has 12 points because it included all the required information shown in the rubric except the graph and an additional point for stating the range.

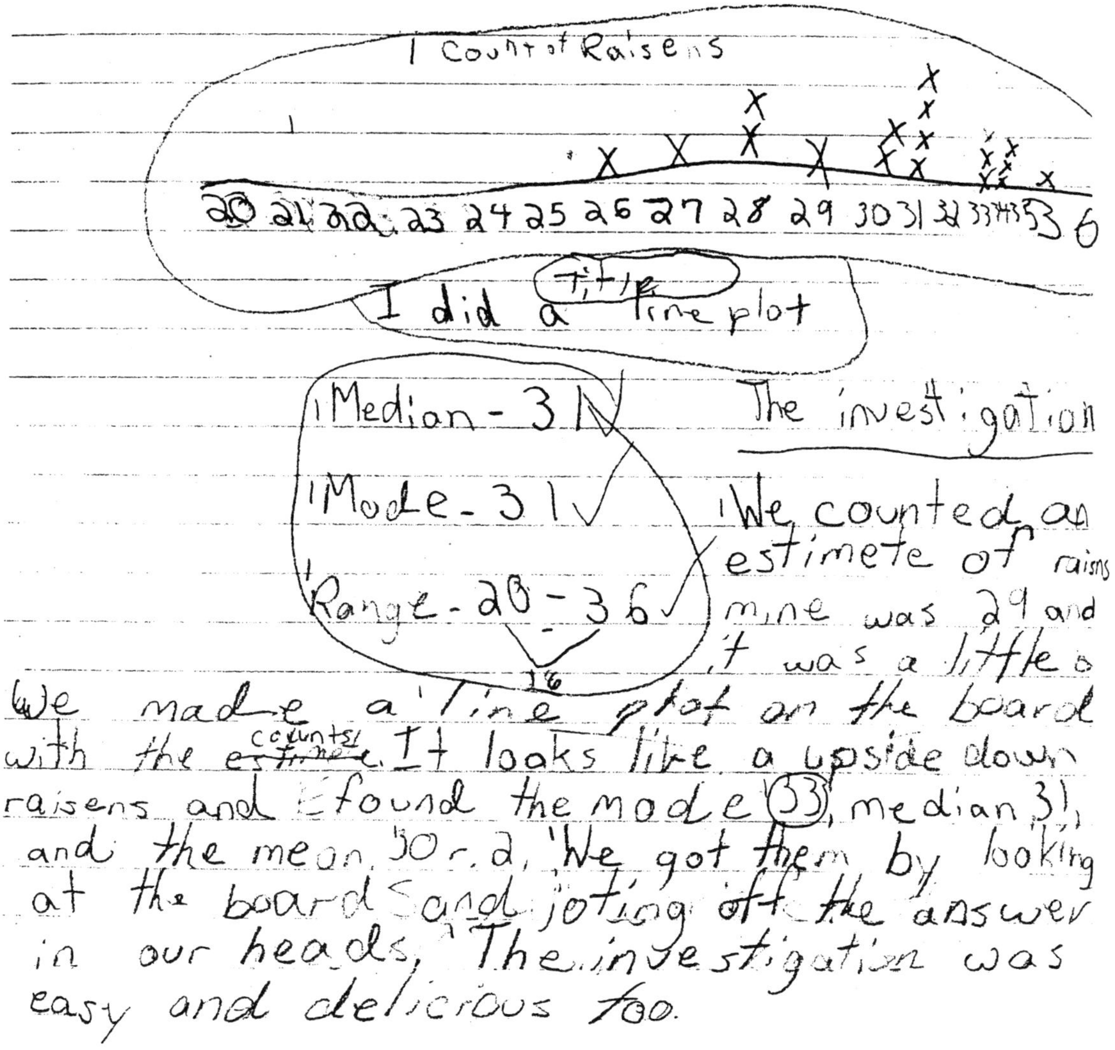

Count of Raisens

I did a Line plot (Title)

Median - 31

Mode - 31 ✓

Range - 20 - 36 ✓

The investigation

We counted an estimete of raisns mine was 29 and it was a little o

We made a line plot on the board with the counts. It looks like a upside down raisens and I found the mode 33, median 31, and the mean 30 r. d. We got them by looking at the boards and joting off the answer in our heads. The investigation was easy and delicious too.

Level 2

The Level 2 report has only five points. There was little explanation about how each of the measures of central tendency was determined. Points were awarded for the graph, the mean, the median, the mode and an explanation of how the investigation was done. There was a copying error showing the mode as 33 in the report, but the correct mode was shown in the chart.

We got a tounce box of sun made raisns.
We got to guess how many raisns there was.
I guessed 12. There was actshuly 26. The range
was 26 ~ 35.

Level 1

The Level 1 report was only given two points, one for the graph and one extra point for the correct range. The student was given the rubric showing only those items checked off, and so was made aware of all the parts that were missing. He was asked to continue to add to the paper until a score of at least eight was reached.

SUMMARY

We have provided a range of questions and classroom scenarios to show how the ideas in this book can be applied to different models. We hope you will try some of them. Be sure to modify the questions and rubrics to meet the needs and interests of your students. Do try to show students a rubric at some point during their work. This, and allowing students to revise their initial efforts, will produce much more robust responses.

You will find additional questions embedded throughout several of the chapters in the book. Look for boxed text. The Appendix also gives photo-copy-ready questions for each of the mathematical strands listed at the start of this chapter.

APPENDIX

STUDENT HANDOUTS

Name: _____

Date: _____

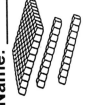

What's It Worth? ~ Base 10

Think of something that you see and/or use in everyday life.
Make a structure that represents this item.
Use some units, some longs, some flats.
Describe using words and/or pictures.

Estimate the value of your structure.
Put a line around it.
Find the actual value.
Tell/show how you arrived at your answer.

Name: _____

Date: _____

Heads and Tails and Other Such Things

We figured out how many **heads**, **tails**, and _____ there were in the problem. This is how we solved it:

Name: _____ **Date:** _____

1. Explain in sentences how prisms are alike and how they are different.

Triangular prisms   Rectangular prisms   Hexagonal prism

2. Explain in sentences how spheres and cylinders are alike and how they are different.

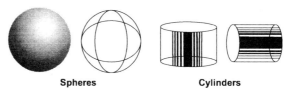

Spheres   Cylinders

Name: _____ Date: _____

POPSICLE PERIMETER

~ Use popsicle sticks.

~ Make a CLOSED SHAPE using all of the sticks.

~ Record how many sticks you used on the graph paper.

~ Record your shape on the graph paper.

~ Make as many different shapes as you can using the same number of popsicle sticks. Record each shape.

~ Write about what you noticed.

Name: _____ **Date:** _____

Perimeter = _____

Name: _____ **Date:** _____

I SPY ...

Describe the picture. Use numbers, words, and/or pictures.

Name: _____ **Date:** _____

ADDING COINS

QUESTION: Mariella just doesn't seem to get the adding of two-digit numbers. Her teacher suggested that she use play money to try to figure out:

$.46 + $.75

Use pictures of coins and describe in words what she needs to do to get a correct solution.

Name: _____ **Date:** _____

Making .25

Beside 1/4, what other fractions will yield .25 on a calculator?
Describe what you notice about how to find more factions.

Name: _____ **Date:** _____

 # TWINS

◆ I made ...

◆ Directions I gave to my partner:

◆ My partner made:

◆ We thought:

Name: _____ **Date:** _____

MEASUREMENT HANDBOOK

Some day we may have visitors from other galaxies!!
It is likely they measure things differently than we do.

- ♦ Use pictures and words to develop a handbook that shows how people on earth measure the length of things that are both short and long.

- ♦ Begin by listing below important words that you are likely to talk about in the handbook, such as ruler, centimeters, length…

- ♦ You can put these in an idea web or in sentences on a new sheet of paper for your first draft.

Name: _____ Date: _____

Raisin Predictions

Look at the raisin data on the board and answer the following questions. Be sure to explain how you arrived at your responses.

What is/are the mode(s)?

What is the median?

If the raisins were collected altogether and redistributed back into the boxes so that each box had the same amount, how many would be in each box?

What trends do you see in the shape of the data?

REFERENCES

Bond, L., E. Roeber, et al. (1997). *Trends in State Student Assessment Programs*. Council of Chief State School Officers.

Bunting, E. (1996). *Market Day*. New York: Joanna Cotler Books.

Clark, E. C. (1999). *I Love You, Blue Kangaroo*. New York: Bantam Doubleday Dell.

Clarke, D. (1988). *Assessment Alternatives in Mathematics*. Melbourne, Australia: Curriculum Corporation.

Collison, J. (1992). "Using Performance Assessment to Determine Mathematical Dispositions." *Arithmetic Teacher, 39*(6), 40–47.

Danielson, C. (1997). *A Collection of Performance Tasks and Rubrics: Upper Elementary School Mathematics*. Larchmont, NY: Eye On Education.

Davis, P. J., and R. Hersh (1981). *The Mathematical Experience*. Boston: Houghton Mifflin.

Dewey, J. (1916). *Democracy in Education*. New York: MacMillan.

Fogarty, R. (1999). "Architects of the Intellect." *Educational Leadership, 57*(3), 76–78.

Freedman, R. L. H. (1994). *Open-Ended Questioning: A Handbook for Educators*. Reading, MA: Addison Wesley.

Herman, J., P. Aschbacher, et al. (1992). *A Practical Guide to Alternative Assessment*. Alexandria, VA: Association for Supervision and Curriculum Development.

Illinois State Board of Education (1995). *Performance Assessment in Mathematics: Approaches to Open-Ended Problems*. Author.

International Association for the Evaluation of Educational Achievement (1997). *TIMSS Mathematics Items: Released Set for Population 1 (Third and Fourth Graders)*. Author.

Kenney, P. A., and E. Silver, Eds. (1997). *Results from the Sixth Mathematics Assessment of the National Assessment of Educational Progress*. Reston, VA: National Council of Teachers of Mathematics.

Kober, N. *Ed Talk: What We Know About Mathematics Teaching and Learning*. Washington, DC: Council for Educational Development and Research.

Lindquist, M. M. (1989). "It's Time to Change." In P. R. Trafton (Ed.), *New Directions for Elementary Mathematics*, pp. 1–13. Reston, VA: National Council of Teachers of Mathematics.

Lindquist, M. M. (1988). *Results from the Fourth Mathematics Assessment of the National Assessment of Educational Progress*. Reston, VA: National Council of Teachers of Mathematics.

Massachusetts Department of Education (1998). *The Massachusetts Comprehensive Assessment System: Release of May 1998 Test Items*. Author.

Mathematical Sciences Education Board (National Research Council) (1993). *Measuring What Counts: A Conceptual Guide for Mathematics Assessment*. Washington, DC: National Academy Press.

Mathematical Sciences Education Board (National Research Council) (1990). *Reshaping School Mathematics: A Philosophy and Framework for Curriculum*. Washington, DC: National Academy Press.

Mathematics Curriculum Framework Development Committee (1997). *Mathematics Curriculum Framework: Achieving Mathematical Power*. Malden, MA: Massachusetts Department of Education.

McKnight, C. C., F. J. Crosswhite, et al. (1987). *The Underachieving Curriculum: Assessing U.S. School Mathematics from an International Perspective*. Champaign, IL: Stipes.

Moon, J., and L. Schulman (1995). *Finding the Connections: Linking Assessment, Instruction and Curriculum*. Portsmouth, NH: Heinemann.

Myren, C. (1995). *Posing Open-Ended Questions in the Primary Classroom*. San Leandro, CA: Teaching Resource Center.

National Commission on Excellence in Education (1983). *A Nation at Risk: The Imperative for Educational Reform*. Washington, DC: US Government Printing Office.

National Council of Supervisors of Mathematics (1996). *Great Tasks and More: A Sourcebook of Camera Ready Resources on Mathematics Assessment*. Golden, CO: Author.

National Council of Teachers of Mathematics (1989). *Curriculum and Evaluation Standards for School Mathematics*. Reston, VA: Author.

National Council of Teachers of Mathematics (1991). *Mathematics Assessment: Myths, Models, Good Questions and Practical Suggestions*. Reston, VA: Author.

National Council of Teachers of Mathematics (1992). *Grade K–6 Addenda Series*. Reston, VA: Author.

National Council of Teachers of Mathematics (1995). *Assessment Standards for School Mathematics*. Reston, VA: Author.

National Research Council (1989). *Everybody Counts: A Report to the Nation on the Future of Mathematics Education*. Washington, DC: National Academy Press.

National Science Board Commission on Precollege Education in Mathematics, Science, and Technology (1983). *Educating Americans for the Twenty-First Century: A Plan of Action for Improving Mathematics, Science, and Technology Education for All American Elementary and Secondary Students So Their Achievement is the Best in the World by 1995*. Washington, DC: US Government Printing Office.

Pandey, T. (1991). *A Sampler of Mathematics Assessment*. Sacramento, CA: California Department of Education.

Rowan, T., and J. Robles (1998). "Using Questions to Help Children Build Mathematical Power." *Teaching Children Mathematics, 4*(9), 504–509.

Stigler, J. W., and J. Hiebert (1997, September). "Understanding and Improving Classroom Mathematics Instruction: An Overview of the TIMMS Video Study." *Phi Delta Kappan*, 14–21.

Tierney, C., and M. Berle-Carman (1998). *Fair Shares*. Columbus, OH: Scott Forseman/Addison Wesley.

TIMS (1997). *Unit Resource Guide II: Parts and Wholes*. Dubuque, Iowa: Kendall/Hunt.

University of Chicago School Mathematics Project (1998). *Grade 3 Teacher's Manual & Lesson Guide*. Chicago: Everyday Learning Corporation.

Wilson, L. (1995). "Enhancing Mathematics Learning with Open-Ended Questions." *Mathematics Teacher, 88*(6), 496–499.

Notes

Notes

Notes